GED Math Practice
2026

Step-by-Step Practice for GED Math Excellence

Dr. Kam Berenji

All inquiries should be addressed to:
szardavai@gmail.com

Welcome to

GED Math Practice

This book is designed to guide you through the exciting and rewarding journey of mastering high school-level mathematics. Whether you're returning to education after some time away or continuing your studies, this book is tailored to help you succeed.

What to Expect:

Inside, you'll find a comprehensive collection of lessons, practice problems, and strategies covering the key topics you'll encounter on the Pre-Calculus exam. We've broken down complex concepts into easy-to-understand sections, ensuring you can follow along and build your skills step-by-step.

How to Use This Book:

1. Start with the Basics: If you're feeling rusty, begin with the foundational topics. Each chapter starts with the basics and gradually progresses to more advanced concepts.

2. Practice, Practice, and Practice: Math is a subject where practice makes perfect. Please take advantage of the numerous practice problems provided and try to solve them without looking at the answers first.

3. Check Your Work: Use the answer keys and explanations to check your work. Understanding your mistakes is crucial for improvement.

4. Utilize the Resources: Don't skip the tips, strategies, and review sections. They are there to help you streamline your study process and give you insights into tackling different types of problems.

Stay Motivated!

Remember, everyone learns at their own pace. It's okay to take your time and revisit challenging sections. With dedication and consistent effort, you'll find yourself improving every day. This book is more than just a study guide—it's your companion in achieving your educational goals.

Good luck, and happy studying!

Contents

List of Figures

List of Tables

CHAPTER # 1

Introduction to GED Math

Math topics that you'll learn in this chapter:

- Importance of the GED Math Test
- Test Structure and Scoring
- Creating a Study Plan
- Effective Study Tips
- Using Practice Tests
- Key Math Concepts Covered
- What You Need to Know for the GED Math Test
- Using a Calculator Effectively

1.1 Understanding the GED

In this section, we delve into the importance of the General Educational Development (GED) credential, focusing specifically on the GED Math Test. We will explore what the GED is, the significance of passing the math test, and provide a detailed overview of the test structure and scoring.

1.1.1 What is the GED?

The GED is a series of high school equivalency exams that allow individuals who did not complete their high school education to demonstrate that they possess the necessary academic skills equivalent to a high school diploma. The test covers four main subjects: Reasoning Through Language Arts, Mathematical Reasoning, Social Studies, and Science.

For many, earning a GED certification is a key milestone in pursuing post-secondary education, qualifying for certain types of employment, or advancing in a career. Partic- ularly, the GED Math Test is a crucial component for those whose educational or career ambitions require a strong foundation in mathematics.

1.1.2 Importance of the GED Math Test

The GED Math Test is designed to assess mathematical skills that are necessary for daily life as well as for post-secondary study. It includes various areas of mathematics, including number operations, algebraic reasoning, measurement, geometry, and data analysis.

A strong performance on the math portion is essential for the following reasons:

- **Post-Secondary Education:** Many college programs require proficiency in mathematics as part of their admissions process.

- **Job Opportunities:** Numerous job roles require a basic understanding of math, and employers may view the GED certification as an indication of critical thinking and problem-solving abilities.

- **Career Advancement:** In some fields, especially in engineering, technology, fi- nance, and healthcare, advanced math skills are necessary for career progression. A solid understanding of GED-level math can lead to more career opportunities.

1.1.3 Test Structure and Scoring

The GED Math Test is structured to evaluate a wide range of essential math skills. Let's break down the organization, topics, and scoring of the test.

Test Format

- The GED Math Test consists of a total of 46 questions.
- You will have 115 minutes (1 hour and 55 minutes) to complete the entire test.
- Test questions may take various forms, such as multiple choice, fill-in-the-blank, drag- and-drop, and drop-down questions.

The use of a calculator is allowed for many of the test questions, but not all of them. Understanding when and how to use the calculator effectively will maximize your time.

Topics Covered:

The primary areas assessed on the test include the following:

- **Number Operations and Number Sense:** Basic arithmetic, fractions, decimals, percentages, and ratios.
- **Algebraic Reasoning:** Understanding and solving equations, inequalities, and systems of equations.
- **Geometry and Measurement:** Properties of geometric shapes, perimeter, area, volume, and coordinate geometry.
- **Data Analysis, Statistics, and Probability:** Interpretation and analysis of data represented in various formats (e.g., graphs, tables) as well as statistics and probability problems.

Sample Problem: Solving Equations

Let's look at a typical problem on the GED Math Test involving solving equations:

EXAMPLE 1

Solve for x in the following equation:

$3x + 7 = 16$

1. We first subtract 7 from both sides:
2. Simplifying
3. Next, divide both sides by 3 to solve for x:

Thus, $x = 3$.

The answer is $x = 3$.

Solution:

$$3x + 7 - 7 = 16 - 7$$

$$\frac{3x}{3} = \frac{9}{3}$$

$$x = \frac{9}{3}$$

$$x = 3$$

Scoring and Passing Standards

Scoring the GED Math Test involves a scaled scoring system ranging from 100 to 200 points. To pass, you must achieve a minimum score of 145 out of 200. However, if you aim for a score of 165 or above, it indicates that you are *college-ready*, making you eligible to bypass certain college placement tests or remedial math classes.

The following table summarizes the score ranges and what they imply:

Score Range	**Performance Description**
100-144	Below Passing Level
145-164	Passing (High School Equivalency)
165-174	College Ready
175-200	College Ready + Credit (Eligible for college credits at some institutions)

Table 1.1: GED Math Test Scoring Breakdown Test Delivery and Scheduling

The GED test is computer-based and offered at official GED testing centers. Many students find it helpful to take several practice tests in a timed setting to get an accurate sense of pacing. Familiarizing yourself with the test's digital environment can significantly reduce test-day anxiety. Here's a useful strategy: break your testing time into thirds. Spend approximately 85 minutes on the questions where you can use a calculator, and leave the remaining 30 minutes for the non-calculator section.

Sample Graph Representation Problem

The GED Math Test often requires you to interpret graphical data. Let's look at an example:

EXAMPLE 2

Question: What is the speed of the object represented by the graph?

1. Use the graph constructed below to answer the question:
2. Since the object covers 6 kilometers in 3 hours, the speed is calculated as:

Thus, the object's speed is 2km/hr.

Solution:

$$\text{Speed} = \frac{Distance}{Time} = \frac{6\ km}{3\ hrs} = 2\frac{km}{hr}$$

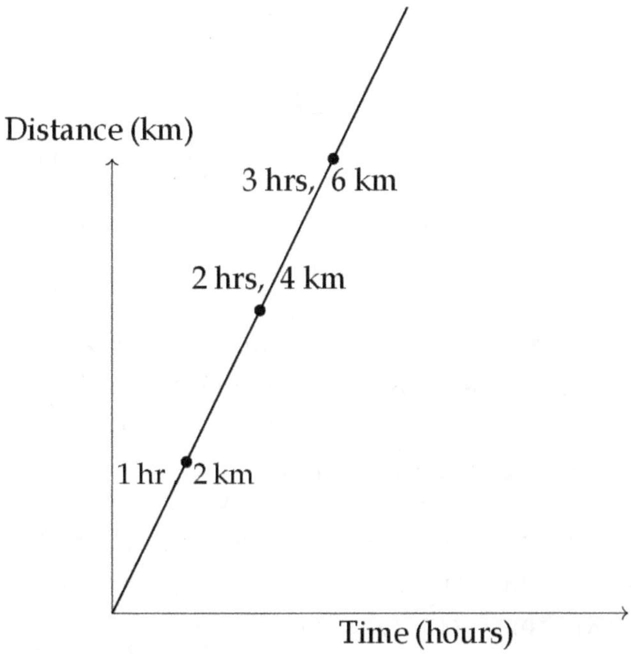

Key Takeaways

Understanding the structure and content of the GED Math Test is essential for building a strong foundation for success. By recognizing the importance of this assessment, familiarizing yourself with the topics covered, and practicing strategies to manage time effectively, you are better prepared to achieve a passing score or even excel beyond the basic requirements. Practice consistently, and focus on areas where you need improvement to boost your confidence and readiness for the test.

Exercises

Exercise 1.1.

1. Conduct research and summarize the historical background of the GED. Highlight any significant reforms or updates since its inception.

2. Discuss why the GED Math Test is an integral part of the certification. What skills does it assess, and why are they important in today's world?

3. Describe the structure of the GED Math Test in detail. Include the types of questions asked and how each section is weighted in the final scoring.

Exercise 1.2.

1. Outline the criteria used to calculate your final GED test score. What is the minimum passing score and what happens if you score below that threshold?

Exercise 1.3.

1. Analyze the significance between the GED test and high school equivalency in both employment and further education. How do they impact your career pathways?

Answers with Explanations

Solution to 1.1:

1. The General Educational Development (GED) test was initially created in 1942. The GED was created by the American Joint Army-Navy Committee to help returning WWII veterans complete high school education. Over time, several content updates have been made to ensure the test aligns with modern educational standards.

2. The GED Math Test covers essential numeracy skills that are critical for various jobs and college-level courses. Topics include arithmetic, algebra, geometry, and data analysis. Mastery of these concepts indicates that a test-taker has the quantitative reasoning skills needed for a majority of modern fields.

3. The GED Math Test is divided into two sections. The first section restricts the use of a calculator, and the second allows it. Questions consist of multiple-choice, drag-and-drop, fill-in-the-blank, and some graphing tasks. Here's a typical score distribution:

Topic	Percentage of Questions
Basic Math (Arithmetic)	20%
Algebra	30%
Geometry	25%
Data Analysis	25%

Table 1.2: GED Math Test Breakdown by Topic

Solution to 1.2:

1. The total score for the GED Math Test is based on the number of correct responses. The passing score for each section of the GED (including Math) is typically around 145 points. If you score below this, you will need to retake that section. High scorers (above 165 points) may qualify for college credit in some institutions.

Solution 1.3:

1. The GED is considered a high school equivalency certification, meaning that it opens up similar education and career pathways to a traditional diploma. Many employers and colleges accept the GED as an indication of academic preparation, though some competitive colleges may prefer a traditional diploma. This is especially important for adults who left high school before completion and want to improve their employment prospects or continue to higher education.

1.2 Study Strategies

Preparing for the GED Math test can seem overwhelming at first, but with effective study strategies and a well-organized plan, it becomes much more manageable. In this section, we will explore techniques to maximize your study efficiency and provide you with the tools needed to succeed. The three key components of any strong preparation regimen should include: creating a study plan, employing effective study techniques, and using practice tests strategically. Let us examine each of these components in detail.

1.2.1 Creating a Study Plan

A study plan is crucial for staying organized and ensuring that you cover all topics before test day. Here's a step-by-step guide to creating an effective study plan:

1. Set Clear Goals

Start by identifying how much time you have remaining before the test date. Divide that time realistically between study sessions, keeping in mind any other commitments you have.

Time Period	Goals	Example
8 Weeks (or more before test)	Review concepts, build strong foundations	Work through arithmetic operations, algebra, and number theory fundamentals.
4–8 Weeks before test	Focus on strengthening weaker areas	If struggling with solving linear equations, spend more time on this section.
2–4 Weeks before test	Begin taking practice tests	Simulate test environments to assess timing and comprehension under test conditions.
Last 2 Weeks	Final review and practice	Review key formulas, solve remaining difficult questions, and focus on test-taking strategies.

The goals and time allotted should be flexible and tailored to your individual needs. For example, if geometry is particularly challenging for you, ensure that you allocate additional time to master those concepts.

2. Break the Content Down

The GED Math test covers several areas: number operations, algebra, geometry, data analysis, and more. Break these larger categories into more manageable subtopics so that you can systematically work through each of them. Consider creating a checklist of topics, where you can mark completed sections.

- For instance, "Number Operations" can be further broken down into:

- Adding and Subtracting Whole Numbers
- Multiplying and Dividing Whole Numbers
- Understanding Fractions, Decimals, and Percents

- "Algebraic Reasoning" would include topics like:

- Expressions and Equations
- Solving Linear Equations
- Understanding Inequalities

Organizing the material this way allows you to focus on mastering one smaller section at a time and to see your progress as you study.

3. Prioritize Weak Areas

Focus on the sections where you feel the least confident. For example, if you find solving inequalities challenging, allocate dedicated study time to review and practice that topic. Conversely, if you are very comfortable solving basic whole-number operations, make that part of your review lighter.

Weak Area Focus Example: If struggling with $ax + b \leq c-d$

Dedicate extra time to reviewing inequality manipulation and hands-on practice by solving multiple variation problems like solving:

$$3x + 5 \leq 20$$

$$x - 7 \geq 2x + 3$$

As you improve, you can shift focus to other areas that need attention.

4. Schedule Regular Breaks

Long, uninterrupted study sessions can lead to burnout. Scheduling short breaks helps maintain focus and sustain productivity.
The graph above visualizes how breaks help reset a learner's focus and improve retention over long periods.

1.2.2 Effective Study Tips

Along with a solid study plan, you'll want to incorporate strategies that help you retain information and get the most out of your study sessions.

1. Take Notes Actively

Passive reading is less effective than engaging with the material. Take notes as you go, summarize key concepts, and mark formulas or examples that you may need to revisit.

For instance, when learning about the quadratic formula, write the formula yourself, then solve several problems to reinforce the concept:

$$x = \frac{-b \pm \sqrt{b^2 + 4ac}}{2a}$$

2. Solve a Variety of Problems

Make sure that you solve a wide variety of problems during your study sessions. This will help ensure that you're familiar with different question types and can apply your knowledge to different contexts.
For example, when learning about fractions, practice:

$$\frac{2}{3} + \frac{5}{6} = \text{Apply LCM: } \frac{4}{6} + \frac{5}{6} = \frac{9}{6} = 1\frac{1}{2}$$

$$\frac{2}{5} x \frac{3}{7} = \text{Multiply Numerators and Denominators: } \frac{6}{35}$$

Solving problems in different formats (word problems, contextual problems) will give you a richer understanding of the concepts.

3. Use Study Tools

Take advantage of tools like calculators, apps, or online exercises to reinforce concepts. When solving equations or learning functions, for example, using a calculator would be essential during the GED Math test.

Calculator Efficiency Tip: Learn to use your calculator in solving multiple-step calculations, such as finding roots, solving a system of equations, or simplifying expressions.

1.2.3 Using Practice Tests

One of the most effective tools for preparing for the GED Math test is taking timed practice tests as they help simulate the test-day environment. Practice tests offer invaluable insights when used correctly.

1. Familiarize with Test Structure

The GED Math test includes a variety of question types, from multiple-choice to fill-in- the-blank and drag-and-drop. Practice tests can acclimate you to the format of these questions. For example, a question might ask you to solve a problem like:

"What is the slope of the line passing through the points $(1, 2)$ and $(3, 4)$?"
You would solve it manually:

$$\text{Slope} = \frac{4-2}{3-1} = 1$$

But practicing it in a timed, test-like format helps accustom you to the pacing needed on test day.

2. Work on Timing and Pacing

Timed practice tests help you assess how much time to spend on each question. If you find yourself spending too long on difficult questions, work on pacing by practicing easier sections first and returning to harder questions using any remaining time.

3. Review Your Mistakes

After completing each practice test, review your mistakes thoroughly. Understand why you got a question wrong and work through similar problems until you're confident with the material. For example, if you missed a question on solving quadratic equations, review the steps for factoring or using the quadratic formula and redo the question.

$$x = \frac{-b \pm \sqrt{b^2 + 4ac}}{2a}$$

Use this as a learning opportunity to refine weak areas.

4. Simulate Test Day Conditions

Finally, when taking practice tests, create conditions as close to test day as possible. Find a quiet room, limit interruptions, and use only the resources allowed during the real GED test (e.g., calculator). This will reduce anxiety on test day and give you confidence in your skills.

Test Condition	Simulated Test Environment	Test Day Benefit
Timed Practice	Complete test in 90 minutes	Build timing strategy for the actual exam
Limited Resources	Use GED-approved calculators	Practical test simulation
Quiet Space	No interruptions	Mimics exam stress conditions

Table 1.3: Simulating Test Day Conditions in Practice Tests

By implementing these comprehensive study strategies, you'll be well-prepared not only to take the GED Math test but to excel in it. Remember, effective preparation is all about planning ahead, using resources wisely, and tackling practice consistently!

Exercises

Exercise 1.4.

1. Outline the steps for creating an effective study plan for the GED Math test. Include how you would allocate time for different sections of the test.

2. Identify and explain three essential study tips that can improve performance in the GED Math test.

3. Discuss how using practice tests can prepare you for the actual GED Math test. Include the benefits of simulating test conditions.

Exercise 1.5.

1. Evaluate the importance of goal-setting in creating an individualized study plan. How does setting specific goals help in tracking progress and staying motivated?

Exercise 1.6.

1. List five online resources or books that provide quality practice tests for the GED

Math exam. Explain why you think these resources are reliable and beneficial.

Answers with Explanations

Solution to 1.4:

1. **Step 1: Assessment** – Determine which sections (Number Operations, Algebra, Geometry, etc.) are your strengths and weaknesses through a diagnostic test.

Step 2: Schedule – Allocate more time to weaknesses while still reviewing strengths. For a two-month plan, you might schedule 30

Step 3: Practice and Review – Incorporate daily practice problems and allocate specific time slots for review sessions. Adjust the plan as necessary based on progress.

2. Three essential study tips:

(a) Spaced repetition – Review materials at increasing intervals to enhance memory retention.

(b) Active recall – Test yourself periodically, rather than passively re-reading notes.

(c) Use a timer – Time yourself as you solve problems to develop speed and accuracy, mimicking the real exam scenario.

3. Practice tests help by:

- Simulating test conditions and timing, reducing test anxiety.
- Identifying areas needing further review.
- Providing insight into the structure of problems and common question formats.

Solution to 1.5:

Goal-setting is crucial in creating an effective study plan as it provides a clear direction. For instance, setting specific goals like "complete 10 practice problems each day" or "achieve 80

As shown in the diagram, the central role of setting goals is keeping on track and motivated throughout the study duration.

Solution to 1.6:

Some reliable resources for GED Math practice tests include:

1. **GED.com**: The official GED site offers authentic practice tests that closely mimic the real exam's format and difficulty level.

2. **Khan Academy**: Offers free, high-quality lessons and practice tests that are aligned with GED Math concepts.

3. **Kaplan's GED Test Prep**: Kaplan's resources are widely regarded and contain comprehensive practice questions and test-taking strategies.

4. **Mometrix Test Preparation**: Provides practice test books and online resources that are regularly updated to reflect the current exam.

5. **ProProfs GED Test Practice**: An online platform that offers a variety of GED practice tests to track progress with instant feedback.

These sources provide varied formats and interactive learning, giving students not just the practice they need, but explanations and feedback for better understanding.

Source Checkpoints:

- Official GED Test Structure
- Proper Level of Difficulty
- Detailed Explanations
- Updated with Recent Exams

The diagram emphasizes the attributes that must be considered when selecting good study resources.

1.3 Math Skills Overview

In this section, we will review the key math skills essential for mastering the GED Math Test. Good preparation requires a deep understanding of core concepts, familiarity with testable material, and efficient use of tools such as a calculator. We will explore:

- Key math concepts covered on the test,
- What you need to know for the GED Math Test, and,
- How to use a calculator effectively during the test.

1.3.1 Key Math Concepts Covered

The GED Math Test assesses a variety of mathematical concepts ranging from basic arithmetic to algebra, geometry, and data analysis. Here are the primary areas covered:

- **Arithmetic**: Operations with whole numbers, fractions, decimals, and percents.
- **Algebra**: Solving linear equations, understanding variables, inequalities, and quadratic equations.
- **Geometry and Measurement**: Identifying geometric shapes, calculating perimeter, area, and volume.
- **Data Analysis**: Understanding data sets, calculating measures of central tendency (mean, median, mode), and recognizing patterns in statistical information.

Let's dive deeper and analyze a few specific examples to clarify these key concepts.

EXAMPLE 1

Basic Arithmetic:
One common operation in basic arithmetic is division. Suppose you're given:

$$\frac{144}{12}$$

The answer is 12, as dividing 144 by 12 yields 12. Division is a crucial skill required across multiple areas, especially when solving real-life problems involving ratios, percents, or unit rates.

EXAMPLE 2

Solving a Linear Equation: You encounter an equation

3x + 5 = 11

1. To solve for x, subtract 5 from both sides of the equation:

2. Next Divide both side by 3

Solution:

$$3x + 5 - 5 = 11 - 5$$

$$\frac{3x}{3} = \frac{6}{3}$$

$$x = 2$$

1.3.2 What You Need to Know for the GED Math Test

The GED Math Test places a heavy focus on three main themes:

1. **Number Operations**: Adding, subtracting, multiplying, and dividing whole numbers, decimals, and fractions.

2. **Algebraic Reasoning**: Translating real-world problems into expressions and equations, solving for unknown values, and interpreting algebraic solutions.

3. **Data Interpretation and Probability**: Analyzing charts, tables, and graphs to extract meaningful data. Set operations and probability experiments are also part of the content.

Key Formulas to Know

While some formulas are provided during the test, it is important to memorize others to save time. Below is a list of essential formulas:

Concepts	**Formula**
Pythagorean Theorem	$a^2 + b^2 = c^2$ (for right triangles)
Area of Circle	$A = \pi r^2$
Slope of a Line	$m = \frac{y_2 - y_1}{x_2 - x_1}$
Quadratic Equation	$ax^2 + bx + c = 0$ (solve using the quadratic formula)

Knowing when and how to apply these formulas can make the critical difference between the correct and incorrect answer on test day.

Example: Finding the Area of a Circle

For a circle with a radius of 5 units, the area is:

$$A = \pi r^2 = \pi \times 5^2 = 25\pi$$

Thus, the area is approximately 78.54 square units when evaluated with $\pi \approx 3.14$.

1.3.3 Using a Calculator Effectively

The GED Math Test allows the use of a scientific calculator for part of the exam. Understanding how to use this tool efficiently can save you valuable time. Here's an overview of how to approach calculator usage.

Functions Available on the Calculator

Most scientific calculators, like the TI-30XS, have a variety of functions:

- **Basic operations**: Addition, subtraction, multiplication, and division.
- **Fractions**: Adding, subtracting, multiplying, and dividing fractions.
- **Exponents and square roots**: Computing powers and roots swiftly.
- **Parentheses**: Grouping terms for correct order of operations.

Important Tip:

Always use parentheses to group terms when performing multiple operations. For example, when solving the following expression:

$$\frac{4+6}{2}x(5^2-1)$$

Entering it into the calculator without grouping as:

$$(4+6) \div 2 \times (5^2 - 1)$$

yields the correct result, 122. Without the parentheses, the calculator result would differ due to order of operations.

When to Use the Calculator

It's crucial to decide when it's better to use the calculator versus doing operations manually. Simple calculations, like 5 × 6, are often quicker to compute mentally than entering the values into the calculator. On the other hand, complex decimal or fraction computations should be left to the calculator to avoid errors.

Example: Using the Calculator for Fractions
Suppose you need to add two fractions:

$$\frac{3}{4}+\frac{5}{8}$$

Without a calculator, you would find a common denominator and then add the numerators. But using the calculator, you can directly input the operation and obtain the result $\frac{11}{8} = 1.375$, which saves you time and minimizes the chance for errors in the computations.

$\frac{n}{d}$ Mode

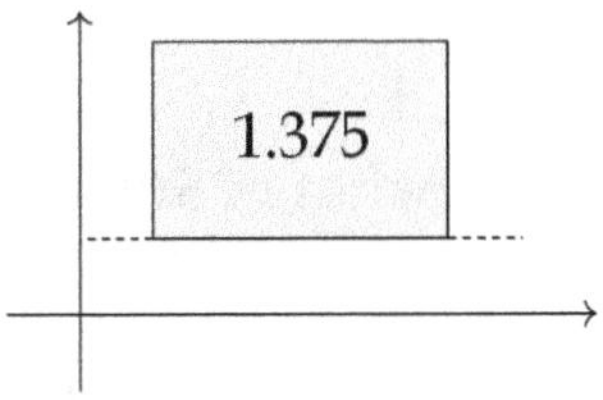

Figure 1.1: Calculator performs fractions instantly and provides decimal equivalence.

Practice With Your Calculator

To ensure you are comfortable with the calculator, practice using it while working through GED-style math problems. Familiarity with its functions and capabilities equips you to confidently approach more complex calculations on the actual test.

Summary

In this section, we've outlined the essential math concepts covered on the GED Math Test, highlighted key areas to focus on, and illustrated how a scientific calculator can assist in making your calculations more efficient. Remember to regularly review formulas, practice core calculations, and become proficient with calculator usage to maximize your performance on the exam. Keep practicing these skills, and you'll strengthen your mathematical foundation in preparation for test day!

Exercises

Exercise 1.7.

1. **List and define at least three key math concepts that are covered in the GED Math Test. Explain why each concept is important for real-life applications.**

2. **Identify and describe three areas of math where calculators are especially useful. Provide an example calculation for each scenario.**

3. **Using the order of operations (PEMDAS), solve the following expression step by step:**

$$7 + (6 \times 5^2 - 12) \div 3$$

Exercise 1.8.

1. **You need to solve this equation using your calculator: 3.45 + 7.89 ÷ 1.23. Explain how to input this into a calculator correctly, following the proper**

order of operations.

2. **Using your calculator, estimate the square root of 500. How does it compare to the exact value that your calculator provides?**

3. **Compute the following using a calculator and check if it follows expectations: 15 mod**

4. **Describe the use of the modulus operator in a real-life context.**

Exercise 1.9.

1. **Without using a calculator, solve the following expression manually:**

$$\frac{15}{4} + (8 \times 4) - 7$$

Then, verify your answer with your calculator.

List two advantages and two disadvantages of relying on calculators during the GED Math Test. Provide specific examples where calculator use could either help or hinder problem-solving.

2. **Explain what you need to be cautious of when using a calculator to solve problems involving exponents, square roots, and fractions.**

Answers with Explanations

Solution to 1.7.

1. Some key math concepts covered in the GED Math Test include:

(a) **Fractions**: Understanding how to add, subtract, multiply, and divide fractions is crucial for everyday applications such as cooking recipes, construction measurements, and budgeting.

(b) **Percentages**: Percentages are commonly used for calculating discounts, taxes, and interest rates, making them vital for financial literacy.

(c) **Algebraic Reasoning**: Algebra helps in forming expressions and solving equations, which is essential in scenarios like determining unknown values in various contexts (e.g., calculating loan payments).

2. Calculators are especially useful in:

(a) Long multiplication and division of decimals: An example would be 153.78 ÷ 2.34.

(b) Square roots: Estimating the square root of non-perfect square numbers such as $\sqrt{17}$.

(c) Exponentiation: Powers of large numbers can be computed effectively, such as 4^5.

3. For the expression:

$$7 + (6 \times 5^2 - 12) \div 3$$

Step 1: Compute the exponent: $5^2 = 25$.

Step 2: $6 \times 25 = 150$.

Step 3: $150 - 12 = 138$.

Step 4: $138 \div 3 = 46$.

Step 5: $7 + 46 = 53$.

Final Answer: 53.

Solution to 1.8.

1. To input $3.45 + 7.89 \div 1.23$ into a calculator, follow the order of operations:

- First, compute $7.89 \div 1.23 = 6.41463$.
- Then, add $3.45 + 6.41463 = 9.86463$.
- Final Answer: 9.86463.

2. Estimating the square root of 500: You can approximate $\sqrt{500}$ to be close to 22.36. Using a calculator, the exact value is approximately 22.36067977. The estimation is quite accurate.

3. The calculation $15 \mod 4 = 3$ since $15 \div 4$ leaves a remainder of 3. In real life, the modulus operation is useful, for instance, in determining cycle repetitions, like how many hours are left after dividing into full-day segments.

Solution to 1.9.

1. Solving the expression manually:

$$\frac{15}{4} + (8 \times 4) - 7$$

Step 1: $15 \div 5 = 3$.
Step 2: $8 \times 4 = 32$.
Step 3: $3 + 32 - 7 = 28$.

The answer is 28. Verifying it with the calculator also gives 28.

2. Advantages of using a calculator:

(a) Speed: Can quickly calculate large numbers, e.g., 234.78×152.39.

(b) Accuracy: Reduces the chance of errors in complex arithmetic like exponentiation.

Disadvantages include:

(a) Over-reliance: Losing your ability to estimate simple calculations such as $15 + 7$.

(b) Input errors: Mistakes in input can lead to wrong solutions, especially if there is confusion with parentheses.

3. When using a calculator for exponents, be conscious of how parentheses are used. For example, -3^2 must be written as $(-3)^2$ if you want the square of -3, not the negative of 3^2. For square roots, ensure that you're applying it entire expressions where required: using $\sqrt{\ }$ improperly can yield incorrect values.

CHAPTER # 2

Number Operations and Number Sense

Math topics that you'll learn in this chapter:

- Basic Arithmetic
- Adding and Subtracting Whole Numbers
- Multiplying and Dividing Whole Numbers
- Order of Operations (PEMDAS)
- Fractions, Decimals, and Percents
- Understanding Fractions
- Converting Between Fractions, Decimals, and Percents
- Operations with Fractions and Decimals
- Ratios and Proportions
- Understanding Proportions
- Solving Proportions
- Applications of Ratios in Real Life
- Positive and Negative Numbers
- Understanding Positive and Negative Numbers
- Adding, Subtracting, Multiplying, and Dividing Negative Numbers
- Key Takeaways

2.1 Basic Arithmetic

In this section, we will delve into essential arithmetic skills that form the foundation of all subsequent mathematical topics. Specifically, we will cover three critical areas of arithmetic: adding and subtracting whole numbers, multiplying and dividing whole numbers, and the order of operations (PEMDAS). Mastering these concepts is crucial for success not only on the GED Math Test but also in numerous real-life applications. Each subsection will include detailed explanations, step-by-step examples, and propositions of useful techniques to build your numeracy skills.

2.1.1 Adding and Subtracting Whole Numbers

Addition and subtraction are the most basic operations in arithmetic, yet they appear in almost all GED Math problems. Whether you are calculating totals, combining quantities, or solving word problems, understanding these two operations is fundamental.

EXAMPLE 1

Simple Addition:
Consider adding the following numbers:

247 + 185

1. To solve this, place the numbers vertically, aligning the digits according to their place values (hundreds, tens, and ones).

2. We add the digits column by column, starting from the right (ones):

7 (ones) + 5 (ones) = 12
(write 2 and carry 1 to the tens column)

4 (tens) + 8 (tens) + 1 (carried over) = 13
(write 3 and carry 1 to the hundreds column)

2 (hundreds) + 1 (hundreds) + 1 (carried over) = 4

Solution:

$$\begin{array}{r} 247 \\ +185 \\ \hline 432 \end{array}$$

7 + 5=12

4 + 8 + 1(carried over) = 13

2 + 1 + 1(carried over) = 4

So, 247 plus 185 is 432

EXAMPLE 2

Simple Subtraction: Let's Subtract:
529 – 364

1. Align the numbers vertically:

2. We subtract digit by digit, beginning from the right (ones). Since 9 is greater than 4, there's no need to borrow in the ones column:

3. For the tens column, 2 - 6 would result in a negative number, so we borrow from the hundreds column, turning the 2 into 12

4. Finally, for the hundreds column:

Solution:

$$\begin{array}{r} 529 \\ -\ 364 \\ \hline 165 \end{array}$$

9 – 4 = 5
12 – 6 = 6
4 – 3 = 1

So, 529 minus 364 is 165.

2.1.2 Multiplying and Dividing Whole Numbers

Multiplication and division are operations used to simplify and break down quantities. They are cornerstones of arithmetic, commonly seen in everyday tasks such as budgeting, cooking, and problem-solving.

Multiplying Whole Numbers

Multiplication can be viewed as repeated addition. For example, calculating 4×3 is the same as adding three 4's together: $4 \times 3 = 4 + 4 + 4 = 12$

Let's look at a multiplication example involving larger numbers

EXAMPLE 1

Multiplication of Two numbers:

Multiply 273 x 8

1. Write 273 on top, and write the number 8 below it.

2. We multiply the digits individually, beginning from the ones.

Solution:

$$\begin{array}{r} 273 \\ \times\ \ 8 \\ \hline 2184 \end{array}$$

$8 \times 3 = 24$ (write down 4 and carry 2)
$8 \times 7 = 56$ (add carried value 2: 56 + 2 = 58) (write down 8 and carry 5)
$8 \times 2 = 16$
(add carried value 5: 16 + 5 = 21)

Hence, $273 \times 8 = 2184$.

Dividing Whole Numbers

Division is the process of distributing a number into equal parts. For example, if you have 12 apples and want to divide them equally among 3 friends, each will get $12 \div 3 = 4$ apples.

EXAMPLE 1

Division of Two Numbers
Divide $426 \div 3$.

1. We use long division in this case. Place 426 under the division symbol, and 3 outside:

2. Start by dividing the hundreds. $4 \div 3 = 1$ with a remainder of 1. Multiply $1 \times 3 = 3$, and subtract $4 - 3 = 1$, then bring down the 2:

3. Bring down the next digit (6):

Solution:

$$\begin{array}{r} 142 \\ 3\overline{)426} \\ -3 \\ \hline 12 \\ -12 \\ \hline 6 \\ -6 \\ \hline 0 \end{array}$$

So, $426 \div 3 = 142$.

2.1.3 Order of Operations (PEMDAS)

To ensure consistency in mathematics, we follow the Order of Operations, abbreviated as PEMDAS:

- **P**arentheses
- **E**xponents
- **M**ultiplication and **D**ivision (from left to right)
- **A**ddition and **S**ubtraction (from left to right)

The order of operations dictates how we solve complex expressions containing multiple operations.

EXAMPLE 1

Using PEMDAS
$6 + 3 \times (4^2 - 2) \div 2$

Following PEMDAS:
1.Solve Parentheses
2. Solve Exponents
Thus, the value of the expression is 27
3. Solve Multiplication and Division (Left to Right)
4. Solve Addition and Subtraction (Left to right)

Solution:

$= 6 + 3 \times (16 - 2) \div 2$; (since $4^2 = 16$)
$= 6 + 3 \times 14 \div 2$; (since $16 - 2 = 14$)
$= 6 + 42 \div 2$; (since $3 \times 14 = 42$)
$= 6 + 21$; (since $42 \div 2 = 21$)
$= 6 + 21 = 27$

Thus, the value of the expression is 27

EXAMPLE 2

Solve the following:
$8 \div 2(2 + 2)$

1. According to PEMDAS, we first resolve the parentheses:

2. Next, we handle the division and multiplication from left to right:

Solution:
$8 \div 2(2 + 2)$

$8 \div 2(4)$

$4(4) = 16$

Thus, the correct result is 16.

Common Pitfall with PEMDAS

A common mistake is to incorrectly treat multiplication as always occurring before division, or addition before subtraction. However, remember that these operations should be performed from left to right, not strictly in the order of "MD" or "AS."

2.1.4 Practice Problems

To reinforce your understanding, try solving the following practice problems:

1. Add: $478 + 329$
2. Subtract: $645 - 278$
3. Multiply: 196×9
4. Divide: $1248 \div 12$
5. Simplify using PEMDAS: $5 + 4 \times (6^2 - 4) \div 2$

2.1.5 Conclusion

In this section, we reviewed the crucial skills of adding, subtracting, multiplying, and dividing whole numbers. Additionally, we introduced the Order of Operations (PEMDAS) as the method for expanding and simplifying mathematical expressions. By mastering these basic arithmetic principles, you will lay a strong foundation for further topics in both algebra and geometry. Completion of the provided practice problems will further enhance your fluency in these operations, helping you to solve a wide range of mathematical problems with accuracy and confidence.

Exercises

Exercise 2.1.

Perform the following additions:

- **368 + 592**
- **845 + 297**
- **1294 + 562 + 438**

Exercise 2.2.

Subtract the following pairs of numbers:

- **876 − 349**
- **1200 − 453**
- **543 − 276 − 135**

Exercise 2.3.

Perform the following multiplications:

- **34 × 28**
- **59 × 45**
- **(17 × 23) × 3**

Exercise 2.4.

Divide the following numbers:

- **816 ÷ 12**
- **965 ÷ 29**
- **(540 ÷ 45) ÷ 2**

Exercise 2.5.

Solve the following expressions using the correct order of operations (PEMDAS):

- **$5+6\times(2+3)$**
- **$50\div5+8\times(6-3)$**
- **$(12+4)^2-3\times5$**

Answers with Explanations

Solution to 2.1. Addition:

- $368+592=960$
- $845+297=1142$
- $1294+562+438=2294$

To illustrate the addition process, let's consider the first problem:

$$\begin{array}{r} 368 \\ +592 \\ \hline 960 \end{array}$$

We add digits from right to left, carrying over as necessary.

Solution to 2.2. Subtraction:

- $876-349=527$
- $1200-453=747$
- $543-276-135=132$

For the second problem, subtract 453 from 1200:

$$\begin{array}{r} 1200 \\ -0453 \\ \hline 0747 \end{array}$$

Notice the borrowing process as you subtract digit by digit.

Solution to 2.3. Multiplication:

- $34\times28=952$
- $59\times45=2655$
- $(17\times23)\times3=1173$

To demonstrate 34×28, first break down the numbers:

$$(30+4) \times (20+8) = 30 \times 20 + 30 \times 8 + 4 \times 20 + 4 \times 8 = 600 + 240 + 80 + 32 = 952$$

Solution to 2.4. Division:

- $816 \div 12 = 68$
- $965 \div 29 = 33.28$
- $(540 \div 45) \div 2 = 6$

Solution to 2.5. Order of Operations (PEMDAS):

- $5 + 6 \times (2+3) = 5 + 6 \times 5 = 5 + 30 = 35$
- $50 \div 5 + 8 \times (6-3) = 10 + 8 \times 3 = 10 + 24 = 34$
- $(12+4)^2 - 3 \times 5 = 16^2 - 15 = 256 - 15 = 241$

To explain the second problem step by step: First, handle the parenthesis: $6 - 3 = 3$. Next, resolve any multiplication and division from left to right:

$$50 \div 5 = 10 \quad , \quad 8 \times 3 = 24$$

Finally, perform the addition: $10 + 24 = 34$

2.2 Fractions, Decimals, and Percents

In this section, we will explore three essential concepts frequently tested on the GED Math Test: fractions, decimals, and percents. These concepts form the bedrock of various mathematical topics and are crucial for solving real-world problems involving measurements, proportions, money, and more. By mastering these topics, you'll be equipped to handle various scenarios that can appear on the GED math exam.

We will cover:

- Understanding Fractions
- Converting Between Fractions, Decimals, and Percents

- Operations with Fractions and Decimals Let's delve into each subsection in detail.

2.2.1 Understanding Fractions

A **fraction** represents the part of a whole. A fraction is written in the form:

$$\frac{a}{b}$$

Where:

- a is the numerator, representing the number of parts being considered.
- b is the denominator, representing the total number of equal parts in the whole.

For example, consider the fraction $\frac{3}{5}$. This means we are considering 3 parts out of a total of 5 equal parts. This can be visualized as follows:

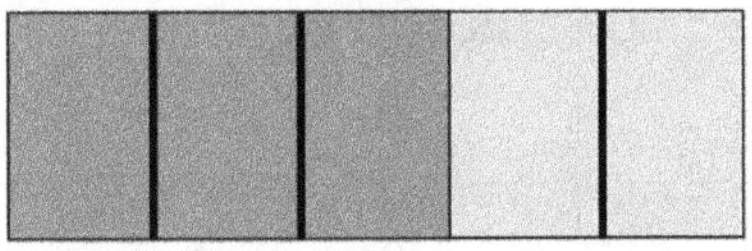

Figure 2.1: Illustration of the fraction $\frac{3}{5}$: Three out of five parts are shaded.

Types of Fractions

Fractions can be classified into different types based on their numerator and denominator:

- **Proper Fraction:** The numerator is less than the denominator, e.g., $\frac{2}{3}$.
- **Improper Fraction:** The numerator is greater than or equal to the denominator, e.g. $\frac{7}{4}$,.
- **Mixed Number:** Contains both a whole number and a fraction, e.g., $2\frac{1}{2}$.

Reducing Fractions

To simplify a fraction, divide both the numerator and denominator by their greatest common divisor (GCD). For example:

$$\frac{12}{16} = \frac{12 \div 4}{16 \div 4} = \frac{3}{4}$$

This means that $\frac{12}{16}$ is equivalent to $\frac{3}{4}$ when reduced.

2.2.2 Converting Between Fractions, Decimals, and Percents

Conversions between fractions, decimals, and percents are essential skills. Here's how you perform each conversion step by step.

Converting Fractions to Decimals

To convert a fraction to a decimal, divide the numerator by the denominator. For example:

$$\frac{3}{4} = 3 \div 4 = 0.75$$

As a result, $\frac{3}{4}$ is equivalent to 0.75 in decimal form.

Converting Decimals to Fractions

To convert a decimal to a fraction, determine the place value of the final decimal digit. Then, place the decimal number over the corresponding power of ten and simplify if necessary. For example, to convert 0.75 into a fraction:

$$0.75 = \frac{75}{100}$$

Now, simplify $\frac{75}{100}$ by dividing both the numerator and denominator by 25:

$$\frac{75 \div 4}{100 \div 4} = \frac{3}{4}$$

Thus, 0.75 is equivalent to the fraction $\frac{3}{4}$

Converting Fractions to Percents

To convert a fraction to a percent, first convert the fraction to a decimal and then multiply it by 100. For example, to express $\frac{2}{5}$ as a percent:

$$\frac{2}{5} = 2 \div 5 = 0.4$$

Thus, $\frac{2}{5}$ is equivalent to 40%

Converting Percents to Fractions

To convert a percent to a fraction, divide the percent by 100 and simplify. For example, to convert 25% to a fraction:

$$25\% = \frac{25}{100} = \frac{1}{4}$$

Visualizing the Relationship Between Fractions, Decimals, and Percents

We can summarize the relationships between fractions, decimals, and percents in the following table:

Fraction	Decimal	Percent
$\frac{1}{4}$	0.25	25%
$\frac{1}{2}$	0.50	50%
$\frac{3}{4}$	0.75	75%
$\frac{1}{5}$	0.20	20%
$\frac{7}{10}$	0.70	70%

Table 2.1: Conversion between fractions, decimals, and percents

2.2.3 Operations with Fractions and Decimals

Now that we understand how to work with fractions and decimals, let's dive into performing operations such as addition, subtraction, multiplication, and division.

Adding and Subtracting Fractions

To add or subtract fractions, the denominators must be the same. If the denominators are different, find the least common denominator (LCD), convert the fractions to have this common denominator, and then perform the operation.
For example:

$$\frac{1}{3} + \frac{2}{5}$$

The least common denominator of 3 and 5 is 15. Convert each fraction:

$$\frac{1}{3} = \frac{5}{15}, \quad \frac{2}{5} = \frac{6}{15}$$

Now add the fractions:

$$\frac{5}{15} + \frac{6}{15} = \frac{11}{15}$$

For a subtraction example:

$$\frac{7}{8} - \frac{1}{4}$$

Convert $\frac{1}{4}$ to have the same denominator as $\frac{7}{8}$:

$$\frac{1}{4} = \frac{2}{8}$$

Now subtract:

$$\frac{7}{8} - \frac{2}{8} = \frac{5}{8}$$

Multiplying Fractions

To multiply fractions, multiply the numerators and multiply the denominators. For example:

$$\frac{2}{5} \times \frac{3}{4} = \frac{2 \times 3}{5 \times 4} = \frac{6}{20}$$

Now simplify the result:

$$\frac{6}{20} = \frac{3}{10}$$

Dividing Fractions

To divide fractions, multiply by the reciprocal of the divisor. The reciprocal of a fraction $\frac{a}{b}$ is $\frac{b}{a}$. For example:

$$\frac{3}{7} \div \frac{2}{5} = \frac{3}{7} \times \frac{5}{2} = \frac{15}{14}$$

Operations with Decimals

Performing operations with decimals is similar to whole numbers, but you must account for the decimal points.

Example 1: Adding Decimals

$$2.34 + 4.56 = 6.90$$

Example 2: Multiplying Decimals

$$3.5 \times 2.7 = 9.45$$

Example 3: Dividing Decimals

To divide decimals, move the decimal point of both numbers to the right until the divisor is a whole number:

$6.72 \div 1.2$ becomes $67.2 \div 12 = 5.6$

2.2.4 Conclusion

Mastering fractions, decimals, and percents is critical for solving mathematical problems involving ratios, proportions, and comparisons. With a solid understanding of how to perform operations across these forms, you'll be well-prepared for GED math questions involving real-world applications.

Exercises

Exercise 2.6.

1. **Simplify the fraction $\frac{15}{35}$**
2. **Convert $\frac{7}{10}$ into a decimal.**
3. **Express 0.75 as a fraction in its simplest form.**

Exercise 2.7.

1. **Convert $\frac{3}{8}$ into a percentage.**
2. **What is 35% as a decimal? Then, express it as a fraction.**
3. **Tim spends $\frac{4}{5}$ of an hour studying math. Convert this to a percentage and a decimal.**

Exercise 2.8.

1. **Add the fractions $\frac{5}{6}+\frac{1}{4}$. Express the result in its simplest form.**
2. **Subtract $\frac{7}{9}-\frac{5}{12}$, showing the steps to find a common denominator.**
3. **Multiply $\frac{2}{3}\times\frac{4}{5}$, and simplify the answer.**

Exercise 2.9.

1. **Divide the fraction $\frac{3}{4}\div\frac{2}{5}$. Write the reciprocal method and simplify the result.**
2. **Convert 0.345 to a fraction, and simplify to the lowest terms.**
3. **Express $1\frac{2}{3}$ as an improper fraction, then subtract $\frac{5}{6}$ from it. Simplify your answer.**

Answers with Explanations

Solution to 2.6.

1. To simplify $\frac{15}{35}$, find the greatest common divisor (GCD) of 15 and 35, which is 5. Dividing both the numerator and denominator by 5 gives:

$$\frac{15}{35} = \frac{15 \div 5}{35 \div 5} = \frac{3}{7}.$$

2. To convert $\frac{7}{10}$ to a decimal, simply divide 7 by 10:

$$\frac{7}{10} = 0.7\,.$$

3. To express 0.75 as a fraction, write it as:

$$0.75 = \frac{75}{100}.$$

Simplifying by dividing both numerator and denominator by 25:

$$\frac{75 \div 25}{100 \div 25} = \frac{3}{4}.$$

Solution to 2.7.

1. To convert $\frac{3}{8}$ into a percentage, divide 3 by 8 then multiply by 100:

$$\frac{3}{8} = 0.375, \quad 0.375 \text{ x } 100 = 37.5\%$$

2. 35% as a decimal is:

$$35\% = \frac{35}{100} = 0.35$$

As a fraction, $\frac{35}{100}$ simplified by dividing both the numerator and the denominator by 5:

$$\frac{35 \div 5}{100 \div 5} = \frac{7}{20}.$$

3. $\frac{4}{5}$ of an hour as a percentage

$$\frac{4}{5} \times 100 = 80\%.$$

As a decimal:

$$\frac{4}{5} = 0.8\,.$$

Solution to 2.8.

1. To add $\frac{5}{6} + \frac{1}{4}$, find the least common denominator (LCD) of 6 and 4, which is 12. Convert each fraction:

$\frac{5}{6} = \frac{10}{12}, \quad \frac{1}{4} = \frac{3}{12}$

Now add:

$$\frac{10}{12} + \frac{3}{12} = \frac{13}{12}, \quad \text{which is } 1\frac{1}{12}.$$

2. To subtract $\frac{7}{9} - \frac{5}{12}$, find the LCD of 9 and 12, which is 36. Convert each fraction:

$$\frac{7}{9} = \frac{28}{36}, \quad \frac{5}{12} = \frac{15}{36}.$$

Now subtract:

$$\frac{28}{36} - \frac{15}{36} = \frac{13}{36}\,.$$

3. To multiply $\frac{2}{3} \times \frac{4}{5}$, multiply both numerators and denominators:

$$\frac{2 \times 4}{3 \times 5} = \frac{8}{15}$$

Solution to 2.9.

1. To divide $\frac{3}{4} \div \frac{2}{5}$ using the reciprocal method, flip the second fraction and multiply:

$$\frac{3}{4} \times \frac{5}{2} = \frac{3 \times 5}{4 \times 2} = \frac{15}{8}.$$

2. To convert 0.345 to a fraction, express it out of 1000:

$$0.345 = \frac{345}{1000}.$$

Simply by dividing both numerator and denominator by 5:

$$\frac{345 \div 5}{1000 \div 5} = \frac{69}{200}.$$

3. First, convert $1\frac{2}{3}$ to an improper fraction:

$$1\frac{2}{3} = \frac{5}{3}.$$

Now subtract $\frac{5}{6}$. First, find the LCD, which is 6.:

$$\frac{5}{3} = \frac{10}{6}, \quad \frac{10}{6} - \frac{5}{6} = \frac{5}{6}.$$

2.3 Ratios and Proportions

In this section, we will dive into the essential concepts of ratios and proportions, which play a vital role not only in mathematics but also in everyday scenarios such as cooking, budgeting, and converting units. Understanding these ideas fully will help you grasp more advanced topics and solve real-world problems with ease. We will start by unpacking the definition of ratios and expand our knowledge to solving proportions and applying these concepts in practical situations.

2.3.1 Understanding Ratios

A *ratio* is a way to compare two or more quantities. It tells us how much of one thing there is compared to another. Ratios can be written in three forms:

1. Using the colon ':' symbol – for example, 2 : 5.
2. In fraction form – for example, $\frac{2}{5}$.
3. In words – for example "2 to 5".

Example 1: Simple Ratio

If in a classroom of 10 students, 6 are boys and 4 are girls, the ratio of boys to girls is 6: 4. This can be simplified (just like simplifying fractions) by dividing both numbers by their greatest common divisor (GCD), which in this case is 2. So, the simplified ratio is:

$$\frac{6}{4}=\frac{3}{2} \text{ or } 3:2$$

This means there are 3 boys for every 2 girls in the classroom.

Example 2: Ratio in Fractional Form

Ratios can also be interpreted as fractions. The ratio 3: 2 can be written as the fraction $\frac{3}{2}$. This is useful when performing calculations where fractional quantities are involved. If for every 3 parts of one substance there are 2 parts of another, we know that the ratio or fraction in which we mix them is $\frac{3}{2}$.

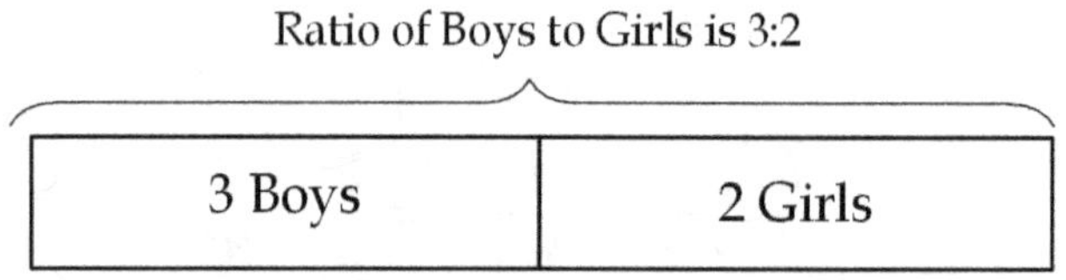

Figure 2.2: Illustration of a 3:2 ratio of boys to girls.

2.3.2 Solving Proportions

A *proportion* is an equation that states that two ratios are equivalent. For example, the statement:

$$\frac{a}{b}=\frac{c}{d}$$

is a proportion, which tells us that the ratio of a to b is equal to the ratio of c to d.

Cross Multiplication in Proportions One of the most common methods for solving proportions is *cross multiplication.* This technique involves multiplying the numerator of one ratio by the denominator of the other and setting both products equal to each other. In the general case, for the proportion:

$$\frac{a}{b}=\frac{c}{d}$$

we 'cross multiply' to get:

$$a \cdot d = b \cdot c$$

This equation can then be solved for the unknown value.

Example 3: Solving a Proportion

Let's solve the proportion:

$$\frac{4}{6} = \frac{x}{9}$$

To solve for x, we cross multiply:

$$4 \cdot 9 = 6 \cdot x$$

That simplifies to:

$$36 = 6x$$

Dividing by 6 on both sides, we find:

$$x = 6$$

Thus, x = 6. This means that if two quantities are in the ratio 4: 6, then x must be 6 when compared to 9 in order to maintain the proportion.

2.3.3 Applications of Ratios in Real Life

The concept of ratios extends beyond classroom problems and is frequently seen in various real-life scenarios. Let's look at a few examples where ratios and proportions play an important role.

Example 4: Cooking Recipes

When you're following a recipe that serves 4 people, but you only need to serve 2 people, you'll need to adjust the quantities of ingredients. By halving the ratios, you can adapt the recipe proportionally.

For example, if the recipe requires:

4 eggs: 2 cups of sugar

To maintain the same taste while serving fewer people, divide both parts of the ratio by 2:

2 eggs: 1 cup of sugar

This keeps the proportion of eggs to sugar the same, but now it's adjusted to serve fewer people.

Example 5: Map Scaling

Maps often use ratios to represent distances. For instance, a map scale might say 1: 50,000, meaning that 1 unit on the map is equivalent to 50,000 units in the real world.

$$\text{Distance on the Map: Real Distance} = 1:50{,}000$$

If a distance of 3 cm is measured on the map, the real-world distance is:

$$3 \times 50{,}000 = 150{,}000 \text{ cm} = 1.5 \text{ km}$$

Thus, the real distance represented by 3 cm on the map is 1.5 kilometers.

Critical Thinking: Recognizing a Misleading Ratio Application Not all usages of ratios are clear-cut! Sometimes ratios can be misrepresented, leading to incorrect conclusions. For example, in advertisements for interest rates or in converting currencies, ratios may be shown in a favorable light by altering the context. Understanding ratios helps avoid these pitfalls.

Scenario	**How Ratios Work**
Cooking	Adjust portions by maintaining ingredient ratios.
Map Reading	Scale distances on a map to real-world equivalents.
Finance (Interest Rates)	Use ratios to understand loan interest over time.

Table 2.2: Common Real-Life Applications of Ratios.

2.3.4 Conclusion

Ratios and proportions are fundamental mathematical concepts that have wide applications in everyday life. Whether you are mixing ingredients, interpreting map distances, or making financial decisions, understanding how to use and manipulate ratios allows you to make accurate calculations and understand relationships between quantities. Remember that solving proportions often involves cross-multiplication, and being able to apply these skills will make many practical problems easier to handle. Keep practicing, and soon, ratios and proportions will become second nature!

Exercises

Exercise 2.10.

You have a bag containing 3 red balls and 5 blue balls.

1. What is the ratio of red balls to blue balls?

2. What is the ratio of red balls to the total number of balls?

3. If 2 additional red balls are added to the bag, what is the new ratio of red balls to blue balls?

Exercise 2.11.

Solve the following proportion:

$$\frac{3}{x} = \frac{9}{12}$$

What is the value of x? Provide a step-by-step explanation.

Exercise 2.12.

In a recipe, the ratio of sugar to flour is 2:3. If you have 600 grams of flour, how much sugar is needed to maintain the ratio?

Answers with Explanations

Solution to 2.10.

1. The ratio of red balls to blue balls is:

$$\frac{3}{5} \text{ or simply } 3 : 5.$$

2. The total number of balls is 3 + 5 = 8, so the ratio of red balls to the total number of balls is:

$$\frac{3}{8} \text{ or } 3 : 8\,.$$

3. After adding 2 more red balls, the total number of red balls becomes 3 + 2 = 5. The new ratio of red balls to blue balls is:

$$\frac{5}{5} = 1 : 1.$$

Solution to 2.11.

To solve the proportion

$$\frac{3}{x} = \frac{9}{12}.$$

we can cross-multiply:

$$3 \times 12 = 9 \times x \quad \Rightarrow \quad 36 = 9x.$$

Solution to 2.12.

The given ratio of sugar to flour is 2:3. This means that for every 3 parts of flour, there are 2 parts of sugar. If you have 600 grams of flour, let x represent the required amount of sugar. Set up the proportion:

$$\frac{2}{3} = \frac{x}{600}.$$

To solve for x, cross multiply:

$$2 \times 600 = 3 \times x \quad \Rightarrow \quad 1200 = 3x.$$

Now divide both sides by 3:

$$x = \frac{1200}{3} = 400.$$

Therefore, you need 400 grams of sugar.

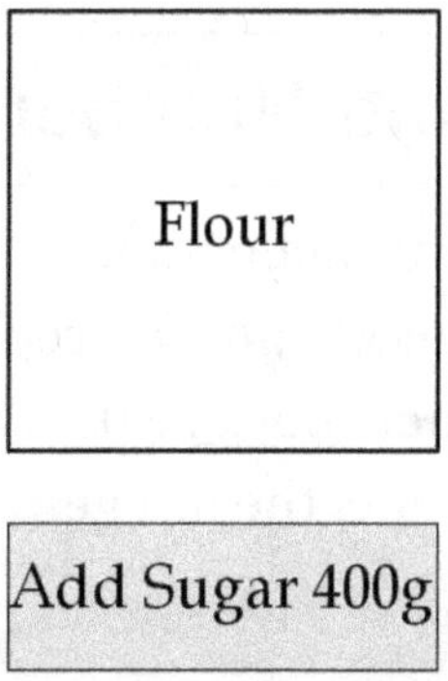

2.4 Positive and Negative Numbers

In this section, we will explore the concept of positive and negative numbers, including how to represent them on number lines, and how to perform arithmetic operations such as addition, subtraction, multiplication, and division. Understanding how to work with these numbers is crucial for success not only in GED math, but also in everyday life.

2.4.1 Understanding Positive and Negative Numbers

In mathematics, numbers can be classified as positive or negative based on their position relative to zero. Positive numbers are greater than zero and appear to the right of zero on a number line. Conversely, negative numbers are less than zero and appear to the left of zero on a number line. The number zero itself is neither positive nor negative; it is neutral.

For example:
- The number +5 is a positive number.
- The number −3 is a negative number.
- The number 0 is neither positive nor negative.

The "+" sign for positive numbers is often omitted, but it's implied. So, 5 is just as positive as +5. Negative numbers, however, always have a "-" sign.

Example 1: Which of the following numbers is positive and which is negative?
−8, 12, 0, −35, 27

- (−8) is a negative number.
- (12) is a positive number.
- (0) is neither positive nor negative.
- (−35) is a negative number.
- (27) is a positive number.

2.4.2 Using Number Lines

A number line is a visual tool that helps us understand the relationship between positive and negative numbers. We can use a number line to represent the size of numbers (their magnitude) and their direction (whether they are positive or negative).
Here is a basic number line that represents the integers from −5 to +5:

Negative Numbers Positive Numbers

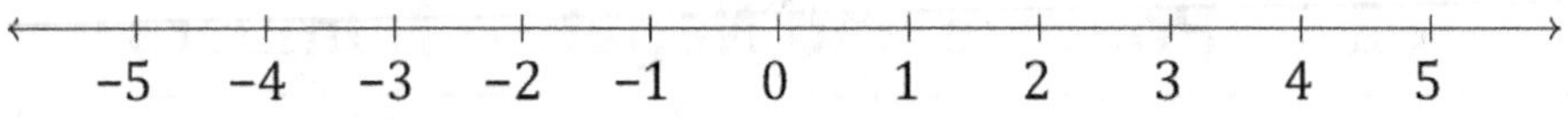

Example 2: On the number line below, locate −4 and 3, and describe where they fall relative to zero:

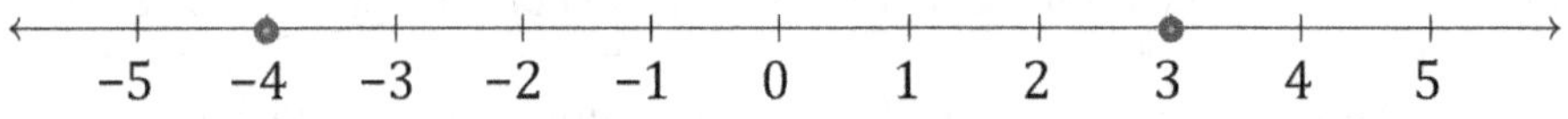

−4 is to the left of zero (it is a negative number). 3 is to the right of zero (it is a positive number).

2.4.3 Adding, Subtracting, Multiplying, and Dividing Negative Numbers

Now that we understand how positive and negative numbers behave on a number line, let's explore how to add, subtract, multiply, and divide them.

Addition and Subtraction of Negative Numbers

When adding or subtracting positive and negative numbers, it can be helpful to think about the number line.

- Adding a positive number moves you to the right.
- Adding a negative number moves you to the left.
- Subtracting a positive number moves you to the left.
- Subtracting a negative number moves you to the right (because subtracting a negative is the same as adding the opposite).

Examples:

1. $5 + (-3)$: Start at 5 on the number line, move 3 units left.

$$5 + (-3) = 2$$

2. $-4 + (-6)$: Start at -4, move 6 units left (because you're adding a negative number).

$$-4 + (-6) = -10$$

3. $-8 - (-2)$: Start at -8, move 2 units right (since subtracting a negative is the same as adding).

$$-8 - (-2) = -6$$

Table: Rules for Addition and Subtraction of Signed Numbers

Operation	Result
Adding two positive numbers $(a + b)$	Always positive. Example: $4 + 6 = 10$
Adding two negative numbers $(a + b)$	Always negative. Example: $-4 + (-6) = -10$
Adding a positive and a negative number	Take the sign of the larger number, and subtract the smaller from the larger. Example: $5 + (-3) = 2$

Subtracting a positive number (a – b)	Move to the left on the number line. Example: 6 – 4 = 2
Subtracting a negative number (a – (−b))	Move to the right on the number line (effectively addition). Example: −3 −(−2) = −3 + 2 = −1

Multiplication and Division of Negative Numbers

The rules for multiplying and dividing positive and negative numbers follow predictable patterns:

- The product of two numbers with the same sign (both positive or both negative) is positive.

- The product of two numbers with different signs (one positive, one negative) is negative.

These rules also apply to division.

Examples:

1. $4 \times (-3) = -12$ (One positive, one negative: the answer is negative).
2. $-5 \times -2 = 10$ (Both negative: the answer is positive).
3. 3. $12 \div (-4) = -3$ (Division follows the same rules as multiplication).
4. 4. $-16 \div -2 = 8$ (Both negative: the answer is positive).

Table: Rules for Multiplication and Division of Signed Numbers

Operation	**Result**
Multiplying two positive numbers ($a \times b$)	Positive. Example: $3 \times 4 = 12$
Multiplying two negative numbers ($a \times b$)	Positive. Example: $-3 \times -4 = 12$
Multiplying a positive and a negative number	Negative. Example: $3 \times -4 = -12$
Dividing two positive numbers	Positive. Example: $12 \div 4 = 3$
Dividing two negative numbers	Positive. Example: $-12 \div -4 = 3$
Dividing a positive by a negative number	Negative. Example: $12 \div -4 = -3$

These rules form the foundation for understanding how to work with negative numbers in all kinds of equations, including those involving algebraic expressions.

2.4.4 Key Takeaways

- Positive numbers are greater than zero, and negative numbers are less than zero.

- A number line is a useful visual tool to understand the relationship between positive and negative numbers.

- To add or subtract signed numbers, follow the rules based on the direction you move on the number line.

- When multiplying or dividing signed numbers, the result depends on whether the signs are the same or different.

By mastering these basic rules for positive and negative numbers, you'll find that other areas in math, such as algebra and geometry, become much easier to navigate. Keep practicing with a variety of problems to build your confidence!

Exercises

Exercise 2.13.

Consider the following numbers: $-5, 3, -1, 7, -10$.

1. **Arrange the numbers in increasing order.**

2. **Identify the number that is closest to zero.**

3. **Calculate the result of $-5 + (-1)$.**

4. **Subtract 7 from -10 and explain your steps.**

Exercise 2.14.

Using the number line provided below, plot the following integers: $-4, 0, 5, -6, 3$. Also, label the midpoints between consecutive integers from -6 to 5.

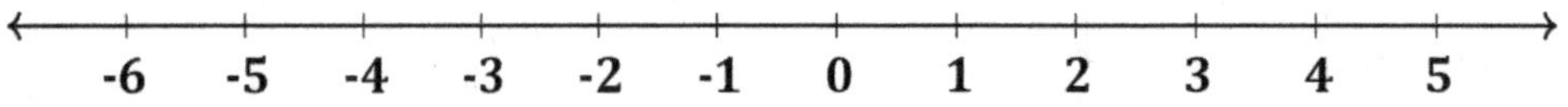

Exercise 2.15.

Consider the expression $12 \times (-3) + (-4) \div 2$.

1. **Solve the multiplication and division parts first (using PEMDAS), showing each step.**

2. **Then complete the sum to find the final answer.**

Exercise 2.16.

Multiply the following pairs of negative numbers and explain the rule used for multiplying negative numbers:

1. $(-7) \times (-2)$

2. $(-4) \times (-5)$

3. **$(-6) \times (-10)$**

Answers with Explanations

Solution to 2.13:

1. Numbers in increasing order: $-10, -5, -1, 3, 7$.

2. The number closest to zero is -1, because its absolute value ($|-1| = 1$) is the smallest.

3. $-5 + (-1) = -6$. Here, when you add two negative numbers, their absolute values are added together and the result is also negative.

4. $-10 - 7 = -17$, because subtracting a positive number from a negative number moves further to the left on the number line.

Solution to 2.14.:

Using the number line:

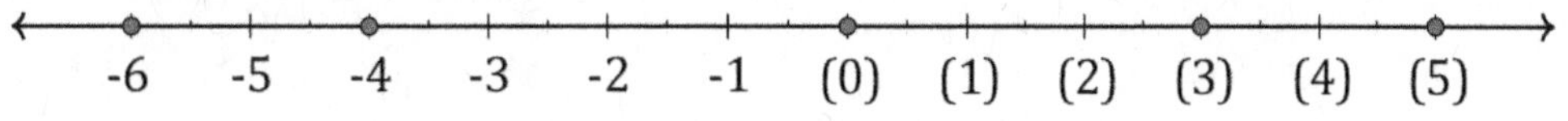

The points -4, 0, 5, -6, 3 are plotted on the number line, and the dashed lines show the midpoints between consecutive integers.

Solution to 2.15.:

Let's break down the expression:

$$12 \times (-3) + (-4) \div 2$$

1. First, solve the multiplication and division as per the order of operations (PEMDAS):

$$12 \times (-3) = -36$$

$$(-4) \div 2 = -2$$

2. Now, sum the results of these operations:

$$-36 + (-2) = -38$$

Thus, the final answer is -38.

Solution 2.16. Exercise 4:

1. $(-7) \times (-2) = 14$. When multiplying two negative numbers, the result is positive.
2. $(-4) \times (-5) = 20$. Again, multiplying two negatives gives a positive result.
3. $(-6) \times (-10) = 60$. The product of two negative values results in a positive.

In general, the rule for multiplying two negative numbers is that the result is always positive.

CHAPTER # 3

Algebraic Reason

Math topics that you'll learn in this chapter:

- Expressions and Equations
- Introduction to Variables
- Writing and Interpreting Algebraic Expressions
- Solving Linear Equations
- Polynomial
- Introduction to Polynomials
- Adding and Subtracting Polynomials
- Multiplying and Dividing Polynomials
- Functions
- Introduction to Functions
- Understanding Input and Output
- Evaluating functions
- Quadratic Equations
- Understanding Quadratic Equations
- Factoring Quadratic Equations

3.1 Expressions and Equations

In this section, we will explore the fundamental building blocks of algebra: expressions and equations. Understanding these concepts is crucial for solving problems in GED Math, especially in the areas of algebraic reasoning. We will cover the basic ideas of variables, how to write and interpret algebraic expressions, and the process of solving linear equations.

Let's dive in and explore each topic with detailed explanations and examples.

3.1.1 Introduction to Variables

A **variable** is simply a symbol—usually a letter—that represents an unknown value or number. In algebra, variables act as placeholders that can take on various numerical values depending on the context of the problem. For instance, if we have an equation like:

$x + 5 = 12$

Here, *x* is a variable that we need to solve for. The goal is to find the value of *x* that makes the equation true.

EXAMPLE 1

Let's solve the equation

$x + 5 = 12.$

1. We want to isolate x by subtracting 5 from both sides:

So, the value of x is 7.

Solution:

$x + 5 - 5 = 12 - 5$

$x = 7$

Real-Life Application:

Suppose you are shopping for books, and each book costs 10. You know that after buying several books and paying 5 in shipping, the total cost is $45. If *x* represents the number of books you bought, the equation would be:

$$10x + 5 = 45$$

You can solve this equation to find out how many books you purchased:

$$10x = 45 - 510x = 40x = \frac{40}{10} = 4$$

So, you bought 4 books.

3.1.2 Writing and Interpreting Algebraic Expressions

An **algebraic expression** is a combination of numbers, variables, and operations (such as addition, subtraction, multiplication, and division) without an equality sign (=).
For example:

$$3x + 2 \text{ or } 7y - 4z$$

are both algebraic expressions. The main goal is to simplify or compute the value of the expression when given the value of the variable(s).

EXAMPLE 2

Consider the expression
$4x + 3$.

If $x = 2$, then we simply substitute $x = 2$ into the expression and compute:

Thus, when $x = 2$, the expression $4x + 3$ evaluates to 11.

Solution:

$$4(2) + 3 = 8 + 3 = 11$$

Writing Expressions from Word Problems:

One of the most valuable skills in algebra is the ability to translate a word problem into an algebraic expression. Let's look at an example:

Example: John earns \$15 per hour. Write an algebraic expression to represent his total earnings after working h hours.

Solution: The total earnings can be represented by the algebraic expression:

$$15h$$

This expression tells us that if John works for h hours, he will earn $15 \times h$ dollars.

3.1.3 Solving Linear Equations

A **linear equation** is an algebraic statement where the highest power of the variable is

1. The general form of a linear equation is:

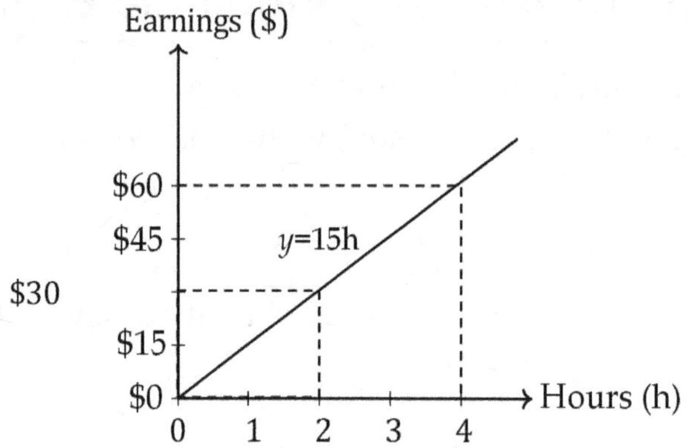

Figure 3.1: Graph representing John's earnings based on hours worked.

$$ax + b = 0$$

Where a and b are real numbers, and x is the variable we need to solve for.

Solving One-Step Equations

In one-step equations, we only need to perform one operation to isolate the variable.

EXAMPLE 3

Solve the equation:

$x - 7 = 10$.

1. To isolate x, we add 7 to both sides:

Solution:

$$x - 7 + 7 = 10 + 7$$

This simplifies to:

$$x = 17$$

Solving Two-Step Equations

In two-step equations, we need to perform two operations to isolate the variable.

EXAMPLE 4

Solve the equation

$3x + 4 = 19$.

Step 1: Subtract 4 from both sides to get rid of the constant term:

Step 2: Divide both sides by 3

Solution:

$$3x + 4 - 4 = 19 - 4$$

This simplifies to:

$$3x = 15$$

$$\frac{3x}{3} = \frac{15}{3}$$

This gives us:

$$x = 5$$

Word Problem Example:
You are designing a garden, and the total cost of the soil and plants is \$200. Each bag of soil costs \$10 and each plant costs \$15. If you buy 10 bags of soil, how many plants can you buy?

Let p represent the number of plants. We set up the equation:

$$10(10) + 15p = 200$$

Simplifying:

$$100 + 15p = 200$$

Subtract 100 from both sides.

$$15p = 100$$

Finally, divide by 15:

$$p = \frac{100}{15} \approx 6.67$$

Since p be a whole number, we conclude that you can buy 6 plants with your budget.

3.1.4 Summary

In this section, we've explored key concepts in algebra including: - Variables, which represent unknown values. - Algebraic expressions, which combine variables, numbers, and operation signs. - The process of solving linear equations, whether one-step or multi-step.

These skills can be applied to a variety of real-world scenarios such as budgeting, shopping, and more complex mathematical reasoning you'll encounter in GED Math. Understanding these concepts will also help you build the foundation necessary to progress towards more advanced topics such as inequalities, functions, and quadratic equations.

The next section will introduce inequalities, where we will study how to solve and graph inequalities on a number line.

Exercises

Exercise 3.1.

1. Define a variable to represent the number of apples in a basket. If 3 apples are added to the basket every minute, write an algebraic expression to represent the number of apples in the basket after t minutes.

2. Given the equation $5x + 3 = 23$, solve for x. Show all steps in your solution.

3. Translate the following statement into an algebraic expression: "Three times the difference of a number and 4 is equal to 18."

Exercise 3.2.

Solve for x in the equation $2(x-1)+4 = 3x-5$.

Exercise 3.3.

Write an algebraic expression for the total cost, C, of purchasing x notebooks at a price of 2.50 each and y pens at 1.20 each.

Answers with Explanations

Solution to 3.1.

1. Let the variable a represent the number of apples currently in the basket. Since 3 apples are added every minute, the total number of apples after t minutes can be expressed as:

$$a(t) = 3t.$$

This algebraic expression shows how the number of apples increases over time.

Solution to 3.1.

2. To solve $5x+3 = 23$, we follow these steps:

- Subtract 3 from both sides:

$$5x = 20$$

- Divide both sides by 5:

$$x = 4$$

Therefore, x is 4

Solution 3.1.

3. The statement "Three times the difference of a number and 4 is equal to 18" can be translated into the following algebraic equation:

$$3(x-4) = 18.$$

To solve for x, proceed as follows:

- Divide both sides by 3:

$$x - 4 = 6$$

- Add 4 to both sides:

$$x = 10$$

Solution 3.2.

To solve $2(x - 1) + 4 = 3x - 5$, follow these steps:

- First, distribute the 2 on the left-hand side:

$$2x - 2 + 4 = 3x - 5.$$

- Simplify the constants on the left:

$$2x + 2 = 3x - 5.$$

- Subtract 2x from both sides:

$$2 = x - 5$$

- Add 5 to both sides:

$$x = 7$$

Therefore, $x = 7$.

Solution 3.3.

To find the total cost:

- The cost of x notebooks at 2.50 each is 2.50x.
- The cost of y pens at 1.20 each is 1.20y.

Therefore, the total cost is given by the algebraic expression:

$$C = 2.50x + 1.20y.$$

Inequalities

Understanding Inequalities

Inequalities are a fundamental topic in algebra that deal with comparisons between two mathematical expressions. Whereas equations express equality, inequalities express relationships where one quantity is either less than or greater than another. In this section, we will explore basic inequalities, their notation, and their meaning.

An "inequality" can take one of four basic forms:

$a < b$	(a is less than b)
$a \leqslant b$	(a is less than or equal to b)
$a > b$	(a is greater than b)
$a \geqslant b$	(a is greater than or equal to b)

Example 1:
Understanding the Symbols Let's start with a simple comparison. Consider the numbers 3 and 7:

$$3 < 7$$

This inequality means that 3 is less than 7.
Now, consider the numbers 5 and 5:

$$5 \leq 5$$

In this case, the inequality $\leq$ implies that 5 is less than or equal to 5. Since the two numbers are equal, the inequality still holds.

Example 2:
Real-Life Application Inequalities are common in real life. For instance, suppose you are driving, and the speed limit is 60 miles per hour. This can be expressed as:

$$v \leq 60$$

where v represents your speed. The inequality indicates that you must drive at a speed less than or equal to 60 mph.

Solving Simple Inequalities

Solving inequalities is quite similar to solving equations, with one important difference: when you multiply or divide both sides of an inequality by a negative number, you must reverse the inequality symbol.

EXAMPLE 3

Solving Linear Inequalities
Solve the inequality:

$$5x - 7 \leq 18$$

Step 1: Add 7 to both sides to isolate the term 5x.

Step 2: Divide both side by 5:

Solution:

$$5x - 7 + 7 \leq 18 + 7$$

$$5x \leq 25$$

$$\frac{5x}{5} \leq \frac{25}{5}$$

$$x \leq 5$$

Final Solution: The solution to the inequality is $x \leq 5$, which means that all values of x less than or equal to 5 will satisfy the inequality.

EXAMPLE 4

Negative Multiplication Consider the inequality:

$-3x > 9$

Step 1: Divide both sides by −3. Remember, since we are dividing by a negative number, we must reverse the inequality symbol:

Solution:

$-3x > 9$

$x < -3$

EXAMPLE 5

Combining Steps:
Solve this inequality step by step:

$2(3x - 4) > 6x + 7$

Step 1: Distribute the 2 on the left-hand side:

Step 2: Subtract 6x from both sides:

Step 3: At this point, you might notice that −8 is never greater than 7. This means there is **no solution** to this inequality.

Solution:

$2(3x - 4) > 6x + 7$

$6x - 6x - 8 > 6x - 6x + 7$

$-8 > 7$

Graphing Solutions on a Number Line

Graphing inequalities is an effective way to visualize all possible solutions. Let's see how we can represent inequalities using number lines.

Graphing $x \leq 5$ To graph the solution $x \leq 5$, we draw a number line and mark a point at 5. Since the inequality includes $x = 5$ (because of the $\leq$ symbol), we use a closed circle at 5 and shade the line to the left to indicate that the solution includes all values less than 5.

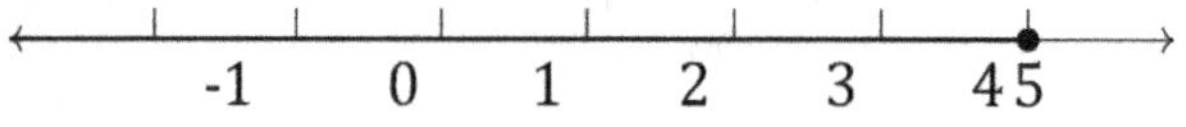

Figure 3.2: Graph of $x \leq 5$.

Graphing $x > -3$ For $x > -3$, we place an open circle at −3 (since x is strictly greater than −3, and −3 itself is not included), and we shade the line to the right.

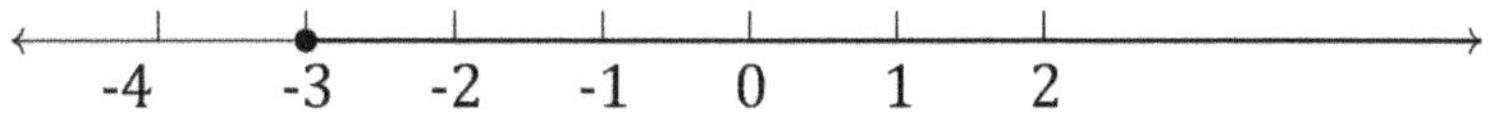

Figure 3.3: Graph of $x > -3$.

Practice Problems:

Here are a few problems for you to test your understanding:

1. Solve for x:

$$4x - 5 > 11$$

2. Solve for y:

$$-2y + 3 \leq 1$$

3. Graph the solution of $x \geq -1$ on the number line.

By regularly practicing these problems and graphing the solutions, you will become proficient in solving and representing inequalities. Keep in mind the core rule: reverse the inequality symbol when multiplying or dividing by a negative number.

Exercises

Exercise 3.4.

Solve the following inequality and express the solution in interval notation: $3x - 5 > 7$

Exercise 3.5.

Solve the compound inequality:

$$-2 \leq 4x + 8 < 16$$

Express the solution in both interval notation and graph it on a number line.

Exercise 3.6.

Determine if the inequality is true or false for $x = -3$:

$$2x + 5 \geq 1 - 3x$$

Exercise 3.7.

A store offers a discount: the price after the discount is less than or equal to 80

Exercise 3.8.

Graph the solution to the inequality:

$$2x - 3 \leq 5$$

on the number line and interpret the result.

Answers with Explanations

Solution 3.4.

The inequality is:

$$3x - 5 > 7$$

We solve it step by step:

$$3x > 7 + 5$$

$$3x > 12$$

$$x > \frac{12}{3} = 4$$

So, the solution is $x > 4$, which in interval notation is:

$$x \in (4, \infty)$$

Graph on a number line:

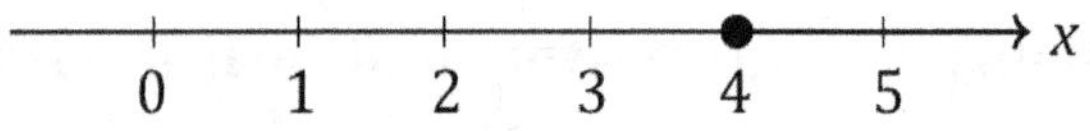

Solution 3.7.

The compound inequality is:

$$-2 \leq 4x + 8 < 16$$

We can break it into two parts. First solve:

$$-2 \leq 4x + 8$$

Subtracting 8 from both sides:

$$-10 \leq 4x$$

Dividing by 4, we get:

$$-\frac{10}{4} \leq x \quad ; \quad x \geq -\frac{5}{2}$$

Now solve the second part:

$$4x + 8 < 16$$

Subtracting 8:

$$4x < 8$$

Dividing by 4:

$$x < 2$$

Hence, the solution in interval notation is:

$$x \in \left[-\frac{5}{2}, 2\right]$$

Graph on a number line:

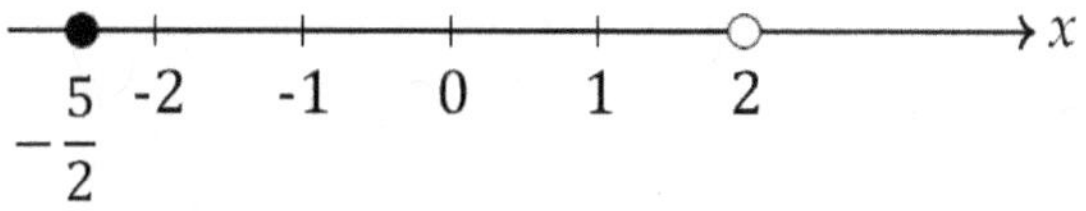

Solution 3.8.

The inequality to check is:

$$2x + 5 \geq 1 - 3x$$

Substituting $x = -3$:

$$2(-3) + 5 \geq 1 - 3(-3)$$

Simplifying:

$$-6 + 5 \geq 1 + 9$$

$$-1 \geq 10$$

This is false. Hence, the inequality is not true for $x = -3$.

Solution 3.9.

We are given that the final price after the discount should be at most 50. This means we can write:

$$0.8p \leq 50$$

Solving for p

$$p \leq \frac{50}{0.8} = 62.5$$

Therefore, the original price of the item must be at most 62.50.

Solution 3.10.

The inequality is:

$$2x - 3 \leq 5$$

Solving:

$$2x \leq 5 + 3$$

$$2x \leq 8$$

$$x \leq \frac{8}{2} = 4$$

So, the solution is $x \leq 4$ or $x \in (-\infty, 4]$.

Graph on a number line:

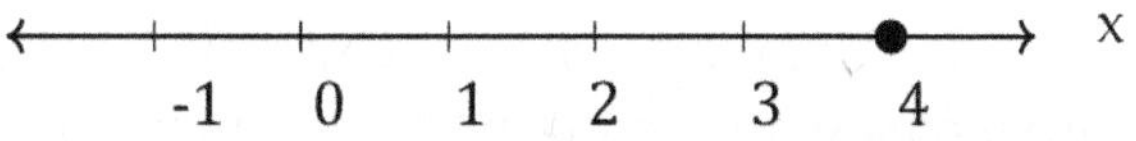

3.2 Polynomials

Polynomials form the foundation of many algebraic operations. In this section, we aim to introduce the concept of polynomials, provide methods for adding and subtracting them, and guide you through multiplying and dividing polynomials. Each step includes examples, strategies, and practice problems, ensuring you feel comfortable working with polynomials.

3.2.1 Introduction to Polynomials

A **polynomial** is an expression that consists of variables (sometimes called indeterminates), constants, and exponents, all combined using the operations of addition, subtraction, and multiplication. A polynomial can have one or more terms, each with a coefficient (the constant multiplied to the variable part).

General Form of a Polynomial:

The general form of a polynomial is:

$$a_n x^n + a_{n-1} x^{n-1} + \cdots + a_1 x + a_0$$

where $a_n, a_{n-1}, \ldots, a_1, a_0$ are the coefficients and x is the variable.

For example, the expression:

$$2x^3 + 5x^2 - 3x + 7$$

is a polynomial of degree 3. The highest power of x (the exponent) is 3, indicating the degree. The term $2x^3$ is called the **leading term**, and 2 is the **leading coefficient**.

Example of a Polynomial:
Consider the polynomial:

$$3x^4 - 2x^3 + 5x - 1$$

- The degree of the polynomial is 4 (the highest exponent).
- The leading coefficient is 3.
- The polynomial contains four terms.

Classification of Polynomials:

Polynomials can be classified based on the number of terms:

- **Monomial**: A polynomial with a single term ($6x^2$).
- **Binomial**: A polynomial with two terms ($2x + 5$).
- **Trinomial**: A polynomial with three terms ($x^2 - 4x + 3$).

3.2.2 Adding and Subtracting Polynomials

When adding or subtracting polynomials, we treat like terms (terms with the same variable raised to the same power) the same way we would add or subtract constants. Only like terms can be combined.

Steps for Addition:

To add polynomials:

1. Align the polynomials so that like terms are in the same column.
2. Add the coefficients of the like terms.

For instance, let's add the following polynomials:

$$(4x^2 + 3x - 5) + (2x^2 - x + 6)$$

Align the like terms:

$$4x^2 + 3x - 5$$

$$2x^2 - x + 6$$

Now combine the like terms

$$(4x^2 + 2x^2) + (3x - x) + (-5 + 6)$$

which simplifies to:

$$6x^2 + 2x + 1$$

Example of Subtraction:

Now let's subtract:

$$(6x^3 + 4x^2 - 5x) - (3x^3 - 2x^2 + 7x - 2)$$

Align the like terms:

$$6x^3 + 4x^2 - 5x + 0$$
$$3x^3 - 4x^2 + 7x - 2$$

Now simplify:

$$(6x\ 3 - 3x\ 3\ ) + (4x\ 2 + 2x\ 2\ ) + (-5x - 7x) + (0 - 2$$

which results in:

$$3x\ 3 + 6x\ 2 - 12x - 2$$

3.2.3 Multiplying and Dividing Polynomials

Multiplication and division of polynomials involve specific techniques that require careful organization. Let's explore these next.

Multiplying Polynomials:

Distributive Property: One of the basic techniques for multiplying polynomials is the distributive property, which states:

$$a(b + c) = ab + ac$$

Extend this to polynomials. Let's multiply:

$$(2x + 3)(x^2 - 4x + 5)$$

Using the distributive property:

$$2x(x^2 - 4x + 5) + 3(x^2 - 4x + 5)$$

Now distribute in each term:

$$2x(x^2) - 2x(4x) + 2x(5) + 3(x^2) - 3(4x) + 3(5)$$

This simplifies to:

$$2x^3 - 8x^2 + 10x + 3x^2 - 12x + 15$$

Now combine like terms:

$$2x^3 - 5x^2 - 2x + 15$$

Multiplying Binomials using FOIL:

A common way to multiply two binomials is the FOIL method, where you multiply:

- **F**irst terms
- **O**uter terms
- **I**nner terms
- **L**ast terms

For example:

$$(x + 2)\ (x - 3)$$

Using FOIL:

$$F : x \times x = x^2$$

$$O : x \times (-3) = -3x$$

$$I : 2 \times x = 2x$$

$$L : 2 \times (-3) = -6$$

Combine the results:

$$x^2 - 3x + 2x - 6 = x^2 - x - 6$$

Dividing Polynomials:

Dividing polynomials follows processes similar to long division.

Let's consider the division of two polynomials:

$$2x^3 - 3x^2 + 4x - 5$$

$$\overline{\qquad\qquad}$$
$$x-1$$

Step 1: Set up the long division

Dividend

$$x-1\overline{\big)\,2x^3-3x^2+4x-5}$$

Step 2: Divide the first term of the dividend by the leading term of the divisor:

$$\frac{2x^3}{x}=2x^2$$

Step 3: Multiply the $2x^2$ by the divisor and subtract from the dividend:

$$(2x^2)(x-1)=2x^3-2x^2$$

$$(2x^3-3x^2)-(2x^3-2x^2)=-x^2$$

Step 4: Bring down the next term and repeat:

$$-(x^2)+4x\div x=-x$$

After following the division steps as you would in numerical long division, you will arrive at the quotient $2x^2-x+3$ with a remainder −8. The final result is:

$$2x^2-x+3-\frac{8}{x-1}$$

Exercises:

1. Add the following polynomials:

$$(3x^3+2x^2)+(4x^3-5x^2+x)$$

2. Subtract:

$$(7x^4+2x^3-4x+5)-(x^4-x^2+6x-8)$$

3. Multiply:

$$(2x+3)\,(x^2-5x+1)$$

4. Divide:

$$\frac{x^3+2x^2-5x+8}{x+2}$$

Answer keys and further explanations for these exercises can be found at the end of the chapter for additional practice.

Exercises

Exercise 3.11.

Simplify the following polynomial expression:

$$(5x^3 - 3x^2 + 7x - 2) - (2x^3 + 4x^2 - 5x + 3)$$

Exercise 3.12.

Add the two polynomials:

$$(2x^2 - 3x + 4) + (x^2 + 5x - 6)$$

Exercise 3.13.

Multiply the following polynomials and simplify:

$$(x - 3)\ (2x^2 + 5x - 4)$$

Exercise 3.14.

Divide the polynomial $6x^3 + 13x^2 - 14x - 8$ by the binomial $x + 2$ using synthetic division or long division. Write your answer in quotient and remainder form.

Exercise 3.15.

Factor the following polynomial completely:

$$2x^3 - 10x^2 + 12x$$

Hint: Start by factoring out the greatest common factor (GCF).

Answers with Explanations

Solution 3.11.

To simplify the expression:

$$(5x^3 - 3x^2 + 7x - 2) - (2x^3 + 4x^2 - 5x + 3),$$

we subtract each corresponding term:

$$(5x^3 - 2x^3) + (-3x^2 - 4x^2) + (7x - (-5x)) + (-2 - 3).$$

Simplifying each term:

$$3x^3 - 7x^2 + 12x - 5.$$

Hence, the final answer is:

$$3x^3 - 7x^2 + 12x - 5.$$

Solution 3.12.

To add the polynomials, combine like terms:

$$(2x^2 - 3x + 4) + (x^2 + 5x - 6).$$

First, group the like terms:

$$(2x^2 + x^2) + (-3x + 5x) + (4 - 6).$$

Now, simplify:

$$3x^2 + 2x - 2.$$

Hence, the final result is:

$$3x^2 + 2x - 2.$$

Solution 3.13.

To multiply $(x - 3)$ and $(2x^2 + 5x - 4)$, use the distributive property:

$$(x - 3)\ (2x^2 + 5x - 4).$$

First, distribute x:

$$x\ (2x^2 + 5x - 4) = 2x^3 + 5x^2 - 4x.$$

Next, distribute −3:

$$-3(2x^2 + 5x - 4) = -6x^2 - 15x + 12.$$

Now add the two expressions:

$$2x^3 + 5x^2 - 4x - 6x^2 - 15x + 12.$$

Group the like terms:

$2x^3 + (-6x^2 + 5x^2) + (-15x - 4x) + 12.$

Simplify:

$$2x^3 - x^2 - 19x + 12.$$

Thus, the final result is:

$$2x^3 - x^2 - 19x + 12.$$

Solution 3.14.

To divide $6x^3 + 13x^2 - 14x - 8$ by $x + 2$ using synthetic division: First, write down the coefficients of the dividend: 6, 13, −14, −8.

-2	6	13	-14	-8
		-12	-2	32
	6	1	-16	24

This means the quotient is:

$$6x^2 + x - 16$$

and the remainder is:

$$24$$

So, the division result is:

$$6x^2 + x - 16 + \frac{24}{x+2}$$

Solution 3.15.

Factor the polynomial

$$2x^3 - 10x^2 + 12x:$$

First, factor out the greatest common factor (GCF), which is 2x:

$$2x\,(x^2 - 5x + 6).$$

Now factor the quadratic $x^2 - 5x + 6$:

$$x^2 - 5x + 6 = (x - 2)\,(x - 3).$$

Therefore, the complete factorization is:

$$2x\,(x - 2)\,(x - 3).$$

3.2.4 Functions

In this section, we will delve into the basics of functions, which are a core concept in algebra. Understanding functions is crucial for solving a variety of algebraic problems, and they frequently appear on the GED Math Test. We will cover the definition and characteristics of functions, how to interpret input and output values of a function, and finally, how to evaluate functions for given inputs.

Introduction to Functions

A **function** is a special kind of relation between two sets: a set of inputs (often called the domain) and a set of outputs (often called the range). What makes a relation a function is that each input is related to exactly one output. To better understand how functions work, let's start with an example.

Example 3.1.

Consider a function $f(x) = 2x + 3$. Here, x is the input, and when you substitute different values for x, you calculate the corresponding output (denoted as f(x)). Let's evaluate this function for several values of x:

$$f(0) = 2(0) + 3 = 3,$$

$$f(1) = 2(1) + 3 = 5,$$

$$f(2) = 2(2) + 3 = 7.$$

In this function, we see that every value of x (our input) is paired with only one value of f(x) (our output). This is what makes $f(x) = 2x + 3$ a function.

Understanding Input and Output

In a function, the set of all possible inputs is known as the **domain**, and the resulting set of outputs is called the **range**. It's important to note that:

- Each input corresponds to exactly one output.
- Different inputs may produce the same output, but a single input cannot be paired with multiple outputs.

Let's visualize this concept using the function f(x) = $\frac{1}{x}$, where the input is any non-zero value since division by zero is undefined. Thus, the domain of this function excludes zero. The following diagram demonstrates the relationship between the inputs and out- puts of a function graphically:

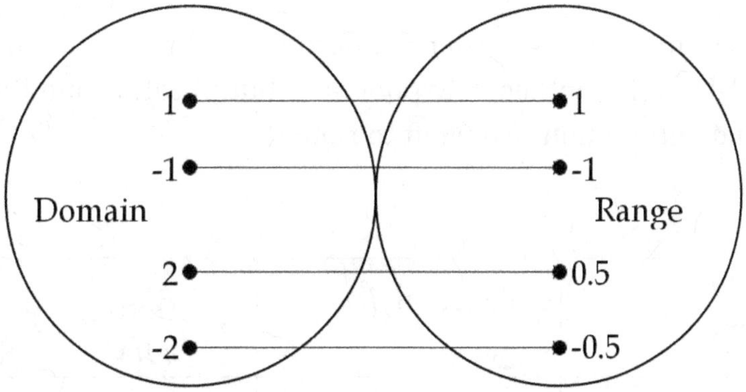

In this diagram, we observe that each input from the domain is paired with exactly one output in the range, which is a key property of functions.

Evaluating Functions

Evaluating a function means finding the output for a specific input. This is done by substituting the input (usually denoted by x) into the function's formula and calculating the result. Below are a few examples to illustrate this concept.

Example 3.2.
Let's evaluate the function $f(x) = 3x^2 - 4x + 2$ for different values of x.

$f(0) = 3(0)^2 - 4(0) + 2 = 2,$

$f(1) = 3(1)^2 - 4(1) + 2 = 1,$

$f(-2) = 3(-2)^2 - 4(-2) + 2 = 12 + 8 + 2 = 22.$

Thus, $f(0) = 2$, $f(1) = 1$, and $f(-2) = 22$.

Example 3.3.

Consider the function $f(x) = x^2 - 5x + 6$. Here's how we evaluate the function for specific values of x:

$$f(1) = 1^2 - 5(1) + 6 = 1 - 5 + 6 = 2,$$

$$f(-3) = (-3)^2 - 5(-3) + 6 = 9 + 15 + 6 = 30,$$

$$f(0) = 0^2 - 5(0) + 6 = 0 + 6 = 6.$$

Thus, the output values are $f(1) = 2$, $f(-3) = 30$, and $f(0) = 6$.

Notice that in each scenario, we substitute the specific value of x into the function and simplify the expression step by step to find the result.

Tip: It's essential to perform correct order of operations when evaluating functions. As a reminder, follow PEMDAS (Parentheses, Exponents, Multiplication and Division (from left to right), Addition and Subtraction (from left to right)).

Input x	Function $f(x) = 2x+3$	Output $f(x)$
0	2(0) + 3 = 3	3
1	2(1) + 3 = 5	5
2	2(2) + 3 = 7	7
-1	2(-1) + 3 = 1	1

Table 3.1: Example of evaluating the function $f(x) = 2x + 3$

Conclusion

Understanding functions is fundamental to algebra, and by mastering the evaluation of functions, you can solve a wide variety of mathematical problems. Remember, a function is a relation where each input is associated with exactly one output, and the process of evaluating a function is simply substituting values into the function's formula and simplifying the resulting expressions.

Through diligent practice, you'll find that working with functions becomes much more intuitive. Keep practicing and refer to examples like those provided here as you work through related questions on the GED Math Test.

Exercises

Exercise 3.14.

1. Define a function in your own words, and provide three real-world examples of functions.

2. Given the following function rule: $f(x) = 3x + 5$, calculate $f(2)$, $f(-1)$, and $f(0)$.

3. Given the set of ordered pairs $\{(1,3), (2,7), (3,11), (4,15)\}$, determine if it represents a function. Justify your answer.

Exercise 3.15.

Given the function $g(x) = x^2 - 4x + 1$:

1. Evaluate $g(0)$, $g(2)$, and $g(3)$.

2. Sketch the graph of this function and identify the vertex.

Exercise 3.16.

Suppose a vending machine is programmed such that the function $c(m) = 2m + 1$ represents the cost of an item, where m is the number of dollars you insert, and $c(m)$ is the total cost in cents.

1. What is the cost of an item when you insert 3,4, and 10? Is the total cost a function of m?

2. If the function were altered to $c(m) = 2m + 2$, what would the new cost be for inserting 1?

Answers with Explanations

Solution 3.14.

1. A function is a relation that assigns each input exactly one output.
Here are three real-world examples:

- The amount of sales tax added to a purchase is a function of the price of the item.
- The distance traveled is a function of speed and time.
- The cost of producing goods is a function of the number of units produced.

2. Let's calculate $f(x) = 3x + 5$ for different values:

$$f(2) = 3(2) + 5 = 6 + 5 = 11$$

$$f(-1) = 3(-1) + 5 = -3 + 5 = 2$$

$$f(0) = 3(0) + 5 = 0 + 5 = 5$$

3. Yes, the set of ordered pairs $\{(1,3),(2,7),(3,11),(4,15)\}$ represents a function because each input value (the first number in the pair) is associated with exactly one output value.

Solution 3.15.

1. First, let's evaluate $g(x) = x^2 - 4x + 1$ at different values of x:

$$g(0) = 0^2 - 4(0) + 1 = 1$$

$$g(2) = 2^2 - 4(2) + 1 = 4 - 8 + 1 = -3$$

$$g(3) = 3^2 - 4(3) + 1 = 9 - 12 + 1 = -2$$

2. The graph of $g(x) = x^2 - 4x + 1$ is shown below. The vertex can be found by using the vertex formula $x = -\frac{b}{2a}$, where a = 1 and b = -4:

$$x = \frac{-(-4)}{2(1)} = \frac{4}{2} = 2$$

Substituting x = 2 into the equation for g(x):

$$g(2) = 2^2 - 4(2) + 1 = -3$$

Therefore, the vertex is (2, −3).

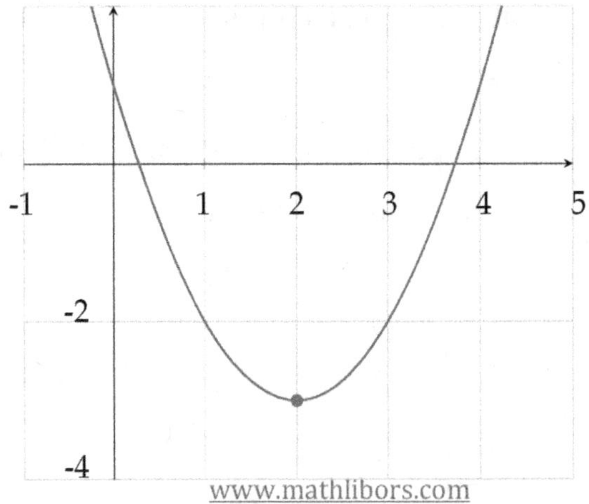

Solution 3.16. For the function $c(m) = 2m + 1$:

1. Let's evaluate the cost of an item for different m values:

$$c(3) = 2(3) + 1 = 7 \text{ cents}$$

$$c(4) = 2(4) + 1 = 9 \text{ cents}$$

$$c(10) = 2(10) + 1 = 21 \text{ cents}$$

Yes, $c(m) = 2m + 1$ is a function because for every value of m (the number of dollars), there is a unique total cost $c(m)$.

2. If the function becomes $c(m) = 2m + 2$, the cost for inserting 1would be:

$$c(1) = 2(1) + 2 = 4 \text{ cents}$$

3.3 Quadratic Equations

Quadratic equations are one of the most important topics in algebra. They are polynomial equations of degree two and play a critical role in various real-world applications, including physics, engineering, and economics. In this section, we'll thoroughly explore quadratic equations, explain techniques for solving them by factoring, and introduce the quadratic formula for cases where factoring may not be possible.

3.3.1 Understanding Quadratic Equations

A quadratic equation is any equation that can be rewritten in the standard form:

$$ax^2 + bx + c = 0$$

In this equation, a, b, and c are constants where $a \neq 0$, and x represents the variable. The term ax^2 makes the equation quadratic (second-degree), while bx is the linear term, and c is the constant term.

Example 1: Standard Form

The following equation is written in quadratic form:

$$3x^2 - 5x + 2 = 0$$

Here:

- $a = 3$, which is the coefficient of x^2 - $b = -5$, which is the coefficient of x - $c = 2$, which is the constant term

Quadratic equations may also occur in factored forms that we'll revisit later, or they may require manipulation (like combining like terms or distributing) to reach the standard form.

Properties of Quadratic Equations

1. Parabolic Shape: If you graph a quadratic function of the form $y = ax^2 + bx + c$, the result is a parabola. The graph opens upwards if $a > 0$ and downward if $a < 0$.

2. Roots/Solutions: The solutions or "roots" of a quadratic equation are the values of x that make $ax^2 + bx + c = 0$. These solutions correspond to the points where the parabola intersects the x-axis.

3. Vertex: The vertex of the parabola is the turning point and can give valuable information about the maximum or minimum value of the function.

Figure 3.4 shows an example of a parabola graphed using the equation $y = x^2 - 4x + 3$.

3.3.2 Factoring Quadratic Expressions

One of the most common ways to solve quadratic equations is by factoring the quadratic expression. Factoring works by rewriting the quadratic expression as a product of two

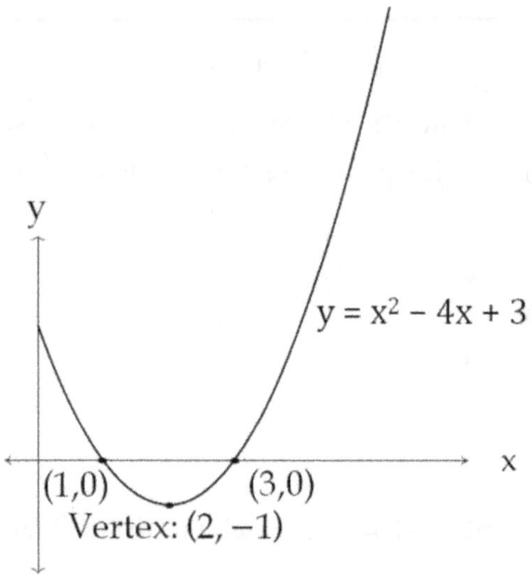

Figure 3.4: Graph of a quadratic function showing roots and vertex.

binomials. This method works best when the coefficients of the quadratic are relatively simple.
The factored form of a quadratic expression looks like this:

$$(ax + m)\ (bx + n) = 0$$

Solving quadratic equations by factoring involves finding the values of x that make each binomial equal to zero. Here's the general procedure:

1. **Ensure the quadratic is in standard form**: $ax^2 + bx + c = 0$
2. **Factor the quadratic expression** (if possible).
3. **Set each factor equal to zero** to solve for x.

EXAMPLE 2

Factoring a Simple Quadratic Consider the quadratic equation:

$$\mathbf{x^2 - 3x + 2 = 0}$$

Step 1: Factor the quadratic expression.

In this case, we look for two numbers that multiply to give 2 (the constant term) and add to give -3 (the coefficient of x). These numbers are -1 and -2.

Step 2: Set each factor equal to zero.

Step 3: Solve for x.

Solution:

$$x^2 - 3x + 2 = 0$$

$x - 1 = 0$ or $x - 2 = 0$

$x = 1$ or $x = 2$

Thus, the solutions to the equation are $x = 1$ and $x = 2$.

Example 3:

Factoring with a Leading Coefficient
Now, let's attempt a slightly more challenging problem with a leading coefficient other than 1:

$$2x^2 - 7x + 3 = 0$$

Step 1: Factor the quadratic expression.

Here, we need two numbers that multiply to $2 \times 3 = 6$ and add to -7. These numbers are -6 and -1, so we rewrite the quadratic as:

$$2x^2 - 6x - x + 3 = 0$$

Group the terms:

$$(2x^2 - 6x) - (x - 3) = 0$$

Factor each group:

$$2x(x - 3) - 1(x - 3) = 0$$

Now factor out the common binomial factor:

$$(x - 3)(2x - 1) = 0$$

Step 2: Set each factor equal to zero.

$$x - 3 = 0 \quad \text{or} \quad 2x - 1 = 0$$

Step 3: Solve for x.

For $x - 3 = 0$, we get $x = 3$. For $2x - 1 = 0$, solving gives $x = \frac{1}{2}$.

Thus, the solutions are $x = 3$ and $x = \frac{1}{2}$.

3.3.3 The Quadratic Formula

Sometimes, factoring may not be possible, at least not through simple methods. In such cases, we turn to the quadratic formula, which provides a systematic way of finding the solutions to any quadratic equation. The quadratic formula is:

$$x = \frac{-b \pm \sqrt{b^2 - 4ac}}{2a}$$

This formula is derived from the process of completing the square and can be used for all quadratic equations, regardless of whether or not they can be factored easily.

EXAMPLE 4

Using the Quadratic Formula
Let's solve the following equation using the quadratic formula:

$$\mathbf{x^2 + 4x + 1 = 0}$$

Step 1: Plug the values of a, b, and c into the quadratic formula

Step 2: Simplify.

Step 3: Simplify further.

Hence, the solutions are:

$x = -2 + \sqrt{3}$

$x = -2 - \sqrt{3}$

Solution:

$$x^2 + 4x + 1 = 0$$

$$x = \frac{-4 \pm \sqrt{4^2 - 4 \cdot 1 \cdot 1}}{2 \cdot 1}$$

$$x = \frac{-4 \pm \sqrt{16 - 4}}{2}$$

$$x = \frac{-4 \pm \sqrt{12}}{2}$$

$$x = \frac{-4 \pm 2\sqrt{3}}{2}$$

$$x = -2 \pm \sqrt{3}$$

EXAMPLE 5

Quadratics with No Real Solutions
Sometimes, a quadratic equation may have no real solutions. For example, consider:

$x^2 + 2x + 5 = 0$

Step 1: Use quadratic formula

Step 2: Simplify

Solution:

$$x^2 + 2x + 5 = 0$$

$$x = \frac{-2 \pm \sqrt{2^2 - 4 \cdot 1 \cdot 5}}{2 \cdot 1}$$

$$x = \frac{-2 \pm \sqrt{4 - 20}}{2}$$

$$x = \frac{-2 \pm 16}{2}$$

Since $\sqrt{16}$ is not a real number, we can conclude that this equation has no real solutions.

3.3.4 Conclusion

In this section, we explored the concept of quadratic equations, broke down methods for solving them via factoring, and introduced the quadratic formula. Whether you're factoring straightforward expressions or dealing with more challenging polynomials, proper understanding of these techniques will empower you to tackle any quadratic equation with confidence, whether by hand or using a calculator. The quadratic formula, in particular, provides a robust solution strategy for all such equations, even when simpler methods like factoring fall short.

Exercises

Exercise 3.17.

Solve the quadratic equation by factoring:

$$x^2 - 5x + 6 = 0$$

1. **Identify the factors of the quadratic expression $x^2 - 5x + 6$.**

2. **Solve for x.**

Exercise 3.18.

Given the quadratic equation:

$$2x^2 - 4x - 6 = 0$$

Solve the equation using the quadratic formula. Include all steps involved in applying the formula:

$$x = \frac{-b \pm \sqrt{b^2 - 4ac}}{2a}$$

where the values of a, b, and care provided by the equation.

Exercise 3.19.

A projectile is launched into the air. The height h (in meters) of the projectile at any time t (in seconds) is modeled by the equation:

$$h(t) = -5t^2 + 20t + 15$$

Find the time at which the projectile reaches its maximum height by completing the square.

Answers with Explanations

Solution 3.19.

To solve $x^2 - 5x + 6 = 0$, we first look for two numbers whose product is 6 and whose sum is -5. The two numbers are -2 and -3. Thus, we factor the quadratic expression:

$$(x - 2)(x - 3) = 0$$

Setting each factor equal to zero, we solve for x:

$$x - 2 = 0 \Rightarrow \quad x = 2$$

$$x - 3 = 0 \quad \Rightarrow \ x = 3$$

Therefore, the solutions are x = 2 and x = 3.

Solution 3.20.

In the quadratic equation $2x^2 - 4x - 6 = 0$, we identify a = 2, b = −4, and c = −6. Substituting these values into the quadratic formula:

$$x = \frac{-(-4) \pm \sqrt{(-4)^2 - 4(2)(-6)}}{2(2)}$$

Step-by-step:

$$x = \frac{4 \pm \sqrt{16 + 48}}{4}$$

$$x = \frac{4 \pm \sqrt{64}}{4}$$

$$x = \frac{4 \pm 8}{4}$$

So, the two solutions are:

$$x = \frac{4 + 8}{4} = 3 \quad and\ x = \frac{4 - 8}{4} - 1$$

Therefore, the solutions are x = 3 and x = −1.

Solution 3.21.

To find the time at which the projectile reaches its maximum height, we start with the given equation:

$$h(t) = -5t^2 + 20t + 15$$

First, we complete the square. Factor out the -5 from the quadratic and linear terms:

$$h(t) = -5(t^2 - 4t) + 15$$

Now, complete the square inside the parentheses. Half of -4 is -2, and squaring it gives 4. Add and subtract 4:

$$h(t) = -5(t^2 - 4t + 4 - 4) + 15$$

$$h(t) = -5((t - 2)^2 - 4) + 15$$

Now, simplify:

$$h(t) = -5(t - 2)^2 + 20 + 15$$

$$h(t) = -5(t - 2)^2 + 35$$

The maximum height is reached when $t - 2 = 0$, or $t = 2$. Thus, the projectile reaches its maximum height at $t = 2$ seconds.

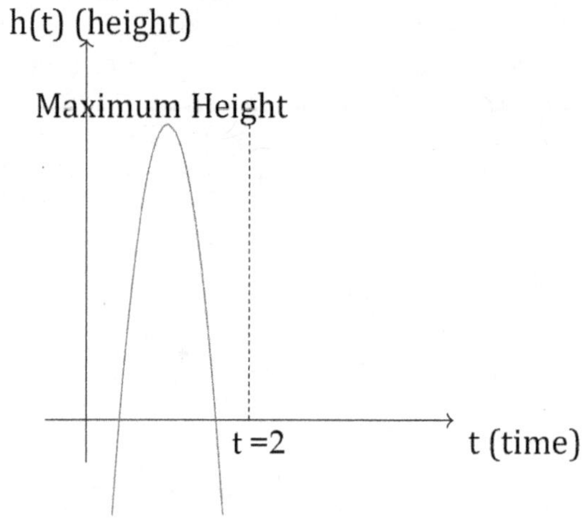

CHAPTER # 3

Geometry and Measurement

Math topics that you'll learn in this chapter:

- Understanding Geometric shapes
- Types of Angles
- Triangles and Their Properties
- Quadrilaterals and Other Polygons
- Perimeter, Area, and Volume
- Calculating Perimeter
- Calculating Area of Various Shapes
- Volume of Solids
- Coordinate Geometry
- The Coordinate Plane
- Plotting Points
- Calculating Slope and Distance Between Points

4.1 Understanding Geometric Shapes

Understanding geometric shapes is a fundamental aspect of geometry and measurement that is essential for mastering the math portion of the GED exam. In this section, we will explore the different types of geometric shapes, focusing on angles, triangles, quadrilaterals, and other polygons. Additionally, we'll provide detailed examples, diagrams, and practical applications to enhance your understanding.

4.1.1 Types of Angles

Angles are an essential building block in geometry. They are formed when two rays meet at a common endpoint, known as the vertex. Let's review the main types of angles based on their degree measures.

- **Acute Angle:** An angle that measures less than 90°.
- **Right Angle:** An angle that measures exactly 90°.
- **Obtuse Angle:** An angle that measures more than 90°, but less than 180°.
- **Straight Angle:** An angle that measures exactly 180°, which looks like a straight line.

Acute Angle $\alpha < 90^\circ$ Right triangle 90° Obtuse Angle $90^\circ < \beta < 180^\circ$

Straight Angle 180°

4.1: Examples of Acute, Right, Obtuse, and Straight Angles.

Example: Identifying Angles in Real Life

Angles appear frequently in everyday life. For example, consider a book standing upright on a table. The angle that the book forms with the surface of the table is a right angle (90°). Another example is the opening of a door. When the door is just slightly ajar, it forms an acute angle with the doorframe. If you open the door halfway, you create a right angle, and continuing to open it creates an obtuse angle.

4.1.2 Triangles and Their Properties

A triangle is a three-sided polygon where the sum of its interior angles is always 180°. There are different types of triangles based on angles and side lengths:

- **Equilateral Triangle:** All three sides and angles are equal. Each angle is 60°.
- **Isosceles Triangle:** Two sides are equal, and the angles opposite these sides are also equal.
- **Scalene Triangle:** All sides and angles are different.
- **Right Triangle:** One of the angles is a right angle (90°).

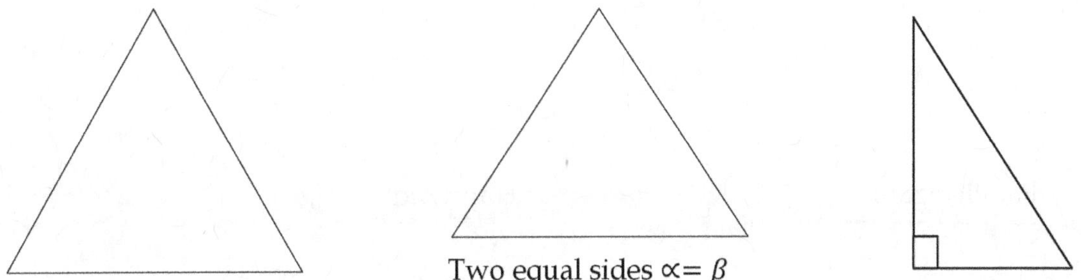

Figure 4.2: Types of Triangles: Equilateral, Isosceles, and Right Triangles.

Example: Calculating the Missing Angle in a Triangle

Let's calculate the missing angle in an isosceles triangle. Suppose two angles in the triangle measure 50° each. We can find the third angle using the fact that the sum of the angles in any triangle is 180°.

$$50^\circ + 50^\circ + \text{Third Angle} = 180^\circ$$
$$100^\circ + \text{Third Angle} = 180^\circ$$
$$\text{Third Angle} = 180^\circ - 100^\circ = 80^\circ$$

Thus, the third angle in this isosceles triangle is 80°.

4.1.3 Quadrilaterals and Other Polygons

A quadrilateral is a polygon with four sides. The sum of the interior angles of any quadrilateral is always 360°. There are several types of quadrilaterals with unique properties:

- **Square:** All sides are equal, and all angles are 90°.
- **Rectangle:** Opposite sides are equal, and all angles are 90°.
- **Rhombus:** All sides are equal, but the angles are not necessarily 90°.
- **Parallelogram:** Opposite sides are equal and parallel, but angles may not be right angles.
- **Trapezoid:** Only one pair of opposite sides is parallel.

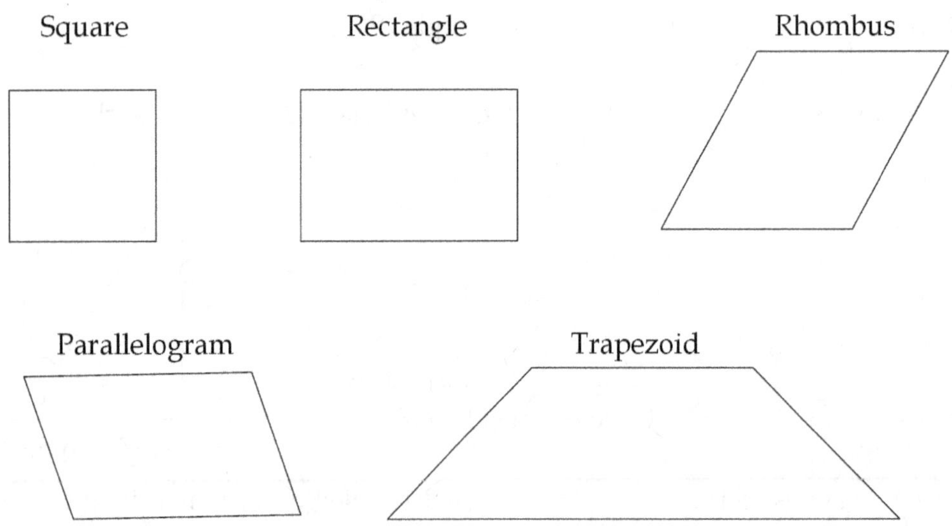

Figure 4.3: Examples of Common Quadrilaterals: Square, Rectangle, Rhombus, Parallelogram, Trapezoid.

Example: Calculating the Area of a Rectangle
The area A of a rectangle can be calculated using the following formula:

$A = \text{Length} \times \text{Width}$
Suppose we have a rectangle with a length of 8 units and a width of 5 units:

A = 8 × 5 = 40 square units

Polygons with More Than Four Sides

Polygons are closed figures with multiple sides. Those with more than four sides are categorized as follows:

- **Pentagon:** A polygon with 5 sides.
- **Hexagon:** A polygon with 6 sides.
- **Heptagon:** A polygon with 7 sides.
- **Octagon:** A polygon with 8 sides.

For any polygon with n sides, the sum of its interior angles can be calculated using the formula:

Sum of interior angles = $180^\circ \times (n - 2)$

For example, the sum of the interior angles of a hexagon ($n = 6$) is:

Sum of interior angles = $180^\circ \times 4 = 720^\circ$

4.1.4 Conclusion

In this section, we covered the basic properties of angles, triangles, quadrilaterals, and other polygons. Understanding these shapes and their attributes is crucial not only for solving geometry problems on the GED exam but also for many real-world applications. Always remember to visualize the shapes and break them down into their individual properties when solving problems. Don't forget to practice with different types of questions to solidify your understanding!

Exercises

Exercise 4.1.

1. Identify the types of angles in the figure below. Label the vertex, and classify each angle as acute, obtuse, right, or straight.

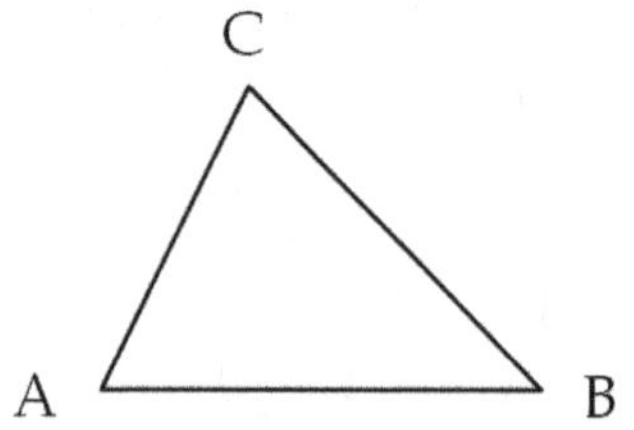

2. **A triangle has angles measuring 40°, 70°, and 70°. What type of triangle is it (equilateral, isosceles, or scalene)? Explain your reasoning.**

3. **For the quadrilateral shown below, find the sum of the interior angles. Then, classify the quadrilateral based on its sides and angles.**

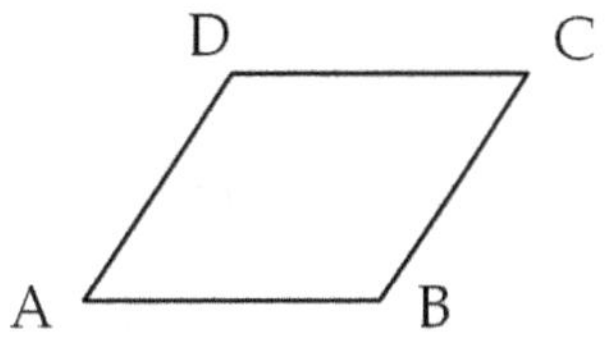

Exercise 4.2.

A right-angled triangle has one leg of 6 cm, and the hypotenuse is 10 cm. Calculate the length of the other leg.

Exercise 4.3.

Draw a polygon with 6 sides, and measure its interior angles using a protractor or calculations. Verify that the sum of the interior angles matches what you expect based on the polygon's number of sides.

Answers with Explanations

Solution 4.1. Exercise 1:

1. In the figure provided, the three angles formed are:

- $\angle A$ is an acute angle ($< 90^\circ$).
- $\angle B$ is a right angle (90°).
- $\angle C$ is an obtuse angle ($> 90^\circ$).

2. The triangle with angles 40°, 70°, and 70° is an **isosceles triangle** because two of its angles are equal.

3. The sum of interior angles for a quadrilateral is 360°. This quadrilateral can be classified as a **trapezoid** because it looks like two sides are parallel but curve slightly apart.

Solution 4.2. Exercise 2:

To solve for the other leg of the right triangle, we use the Pythagorean theorem:

$$a^2 + b^2 = c^2$$

where a = 6, c = 10 is the hypotenuse, and b is the unknown leg. Substituting the values:

$$6^2 + b^2 = 10^2 36 + b^2 = 100 b^2 = 100 - 36 = 64 b = \sqrt{64} = 8 \text{ cm}.$$

Thus, the length of the other leg is **8 cm**.

Solution 4.3. Exercise 3: The sum of the interior angles of any polygon with n sides is given by the formula:

$$(n-2) \times 180^\circ$$

For a polygon with 6 sides (hexagon):

$$\text{Sum of angles} = (6-2) \times 180^\circ = 4 \times 180^\circ = 720^\circ$$

After drawing the polygon and measuring each angle, we verify that the total sum of the interior angles is indeed 720°. This confirms our calculation for the hexagon.

4.2 Perimeter, Area, and Volume

In this section, we will cover how to calculate the perimeter, area, and volume of various geometric shapes. These concepts are widely used in geometry and everyday situations, such as determining the size of a plot of land or the amount of space inside a box. We'll go step by step through each concept, offering clear explanations and multiple examples to help you master these essential geometric measurements.

4.2.1 Calculating Perimeter

The perimeter of a geometric shape is the total length of its outer boundary. For simple polygons, it is the sum of the lengths of all sides. The formula you use for the perimeter depends on the shape in question. Below are common formulas for various shapes.
Rectangle: The perimeter P of a rectangle is given by the formula:

$$P = 2 \cdot (l + w)$$

where l is the length and w is the width of the rectangle.

Example 4.1.
Consider a rectangle with a length of 8 meters and a width of 5 meters. The perimeter is:

$$P = 2 \cdot (8\,m + 5\,m) = 2 \cdot 13\,m = 26\,m.$$

So, the total perimeter is 26 meters.

Triangle: The perimeter P of a triangle is the sum of the lengths of its three sides:

$$P = a + b + c$$

where a, b, and c are the lengths of the sides.

Example 4.2.
Suppose a triangle has sides of 7 cm, 9 cm, and 12 cm. The perimeter would be:

$P = 7\text{ cm} + 9\text{ cm} + 12\text{ cm} = 28\text{ cm}.$

Thus, the perimeter of the triangle is 28 cm.

Circumference of a Circle: The perimeter of a circle is called the *circumference*, and it is given by:

$$C = 2\pi r$$

where r is the radius of the circle.

Example 4.3. If a circle has a radius of 6 cm, the circumference is:

$$C = 2\pi \cdot 6\text{ cm} = 12\pi\text{ cm} \approx 37.7\text{ cm}.$$

So, the circumference is approximately 37.7 cm.

4.2.2 Calculating Area of Various Shapes

The area of a shape is the amount of space it occupies in a plane. Different shapes have different area formulas which are detailed below.
Rectangle: The area A of a rectangle is given by:

$$A = l \cdot w$$

where l is the length and w is the width.

Example 4.4.
Consider a rectangular garden that is 10 meters long and 6 meters wide. The area is:

$$A = 10\text{ m} \times 6\text{ m} = 60\text{ m}^2.$$

Thus, the area of the garden is 60 square meters.

Triangle: The area A of a triangle is:

$$A = \frac{1}{2} b \cdot h$$

where b is the base, and h is the height of the triangle.

Example 4.5. For a triangle with a base of 8 cm and a height of 5 cm, the area is:

$$A = \frac{1}{2} \cdot 8 \text{ cm} \cdot 5 \text{ cm} = 20 \text{ cm}^2.$$

Hence, the area is 20 square centimeters.

Circle: The area A of a circle is calculated as:

$$A = \pi r^2$$

where r is the radius.

Example 4.6. If a pizza has a radius of 4 inches, the area of the pizza is:

$$A = \pi \cdot 4^2 \text{ in}^2 = 16\pi \text{ in}^2 \approx 50.27 \text{ in}^2.$$

Thus, the area of the pizza is approximately 50.27 square inches.

4.2.3 Volume of Solids

Volume refers to the amount of space occupied by a 3D object. Below, we discuss the formulas for the volume of common three-dimensional figures.

Rectangular Prism (Box): The volume V of a rectangular prism is given by:

$$V = l \cdot w \cdot h$$

where l is the length, w is the width, and h is the height.

Example 4.7.
A rectangular box has dimensions 4 cm by 3 cm by 2 cm. The volume of the box is:

$$V = 4 \text{ cm} \times 3 \text{ cm} \times 2 \text{ cm} = 24 \text{ cm}^3.$$

So, the total volume is 24 cubic centimeters.

Cylinder: The volume V of a cylinder is calculated using the formula:

$$V = \pi r^2 h$$

where r is the radius of the base, and h is the height of the cylinder.

Example 4.8.
Imagine a cylindrical can with a radius of 3 cm and a height of 10 cm. The volume is:

$$V = \pi \cdot 3^2 \cdot 10 \text{ cm}^3 = 90\pi \text{ cm}^3 \approx 282.74 \text{ cm}^3.$$

Thus, the volume of the can is approximately 282.74 cubic centimeters.

Sphere: The volume V of a sphere is given by:

$$\frac{4}{3}\pi r^3$$

Where r is the radius.

Example 4.9.
If a sphere has a radius of 5 inches, the volume of the sphere is:

$$V = \frac{4}{3}\pi \cdot 5^3 in^3 = \frac{4}{3}\pi \cdot 125\ in^3 = \frac{500}{3}\pi in^3 \approx 523.6\ in^3$$

Hence, the volume of the sphere is approximately 523.6 cubic inches.

In summary, mastering the concepts of perimeter, area, and volume allows you to efficiently measure planes and spaces across a variety of shapes. Whether you are calculating the fence needed for a garden, determining the space inside a room, or handling more complex 3D shapes, these basic formulas and methods provide the tools you need for accurate geometric analysis.

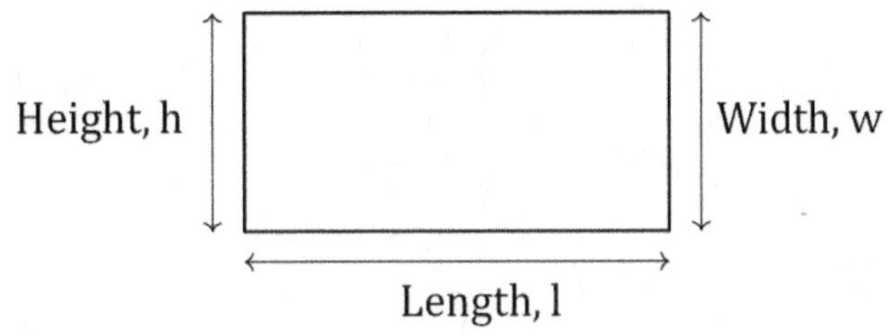

Figure 4.4: Dimensions of a Rectangular Prism

Exercises

Exercise 4.4.

1. **Calculate the perimeter of a rectangle with a length of 8 cm and a width of 5 cm.**

2. **Find the perimeter of an equilateral triangle where each side measures 6.5 cm.**

3. **A circular fence encloses a garden with a radius of 4 meters. Calculate the perimeter (also known as the circumference) of the fence. Use $\pi \approx 3.14$.**

Exercise 4.5.

1. **Calculate the area of a triangle with a base of 10 cm and a height of 7 cm.**

2. **Find the area of a square where each side is 12 cm long.**

3. **A parallelogram has a base of 14 cm and a height of 9 cm. Calculate its area.**

Exercise 4.6.

1. Calculate the volume of a rectangular prism with dimensions 4 cm by 5 cm by 6 cm.

2. Find the volume of a cylinder with a radius of 3 cm and a height of 8 cm. Use $\pi \approx 3.14$.

3. A cone has a radius of 2 cm and a height of 7 cm. Calculate its volume. Use $\pi \approx 3.14$.

Answers with Explanations

Solution 4.4.

1. The formula for the perimeter of a rectangle is $P = 2(l + w)$, where l is the length and w is the width. Substituting the given numbers:

$$P = 2(8\text{ cm} + 5\text{ cm}) = 2 \times 13\text{ cm} = 26\text{ cm}.$$

2. The perimeter of an equilateral triangle is the sum of all its sides. Since each side has the same length:

$$P = 3 \times 6.5\text{ cm} = 19.5\text{ cm}.$$

3. The perimeter (circumference) of a circle is given by the formula $C = 2\pi r$, where r is the radius. Using the approximation $\pi \approx 3.14$:

$$C = 2 \times 3.14 \times 4\text{ m} = 25.12\text{ m}.$$

Solution 4.5

1. The area of the triangle is calculated by the formula $A = \frac{1}{2} \times base \times height$. Substituting the given numbers:

$$A = \frac{1}{2} \times 10cm \times 7cm = 35cm^2$$

2. The area of a square is given by $A = s^2$, where s is the length of one side:

$$A = 12^2 cm^2 = 144cm^2$$

3. The area of a parallelogram is A = base × height. Substituting the given values:

$$A = 14 \text{ cm} \times 9 \text{ cm} = 126\ cm^2 .$$

Solution 4.6.

1. The volume of a rectangular prism is calculated by the formula
V = l × w × h, where l, w, and h are the length, width, and height respectively:

$$V = 4 \text{ cm} \times 5 \text{ cm} \times 6 \text{ cm} = 120 \text{ cm}^3.$$

2. The volume of a cylinder is given by $V = \pi r^2 h$, where r is the radius and is the height. Using $\pi \approx 3.14$:

$$V = 3.14 \times (3 \text{ cm})^2 \times 8 \text{ cm} = 3.14 \times 9 \text{ cm}^2 \times 8 \text{ cm} = 226.08 \text{ cm}^3.$$

3. The volume of a cone is given by $V = \frac{1}{3}\pi r^2 h$. Substituting the given numbers and using $\pi \approx 3.14$:

$$V = \frac{1}{3} \times 3.14 \times (2cm)^2 \times 7cm = \frac{1}{3} \times 3.14 \times 4cm^2 \times 7cm = 29.32cm^2$$

4.2.4 Coordinate Geometry

In this section, we will explore the fundamental concepts of coordinate geometry, which is an essential part of the GED Math test. Understanding how to navigate and interpret the coordinate plane is crucial, as it provides the foundation for many algebraic and geometric concepts. We will cover three key subsections: the coordinate plane, plotting points, and calculating slope as well as the distance between points. Each of these skills will help you solve problems that involve graphs and spatial relationships, both of which are tested extensively.

4.2.5 The Coordinate Plane

The coordinate plane is a two-dimensional plane divided by two perpendicular number lines: the x-axis (horizontal) and the y-axis (vertical). These axes intersect at a point called the "origin", which has coordinates (0, 0).

Each point on the plane can be described by an ordered pair (x,y), where:

- x represents the horizontal distance from the origin,
- y represents the vertical distance from the origin.

The plane is divided into four quadrants:

1. Quadrant I: Both x and y are positive (+, +).
2. Quadrant II: x is negative and y is positive (−, +).
3. Quadrant III: Both x and y are negative (−, −).
4. Quadrant IV: x is positive and y is negative (+, −)

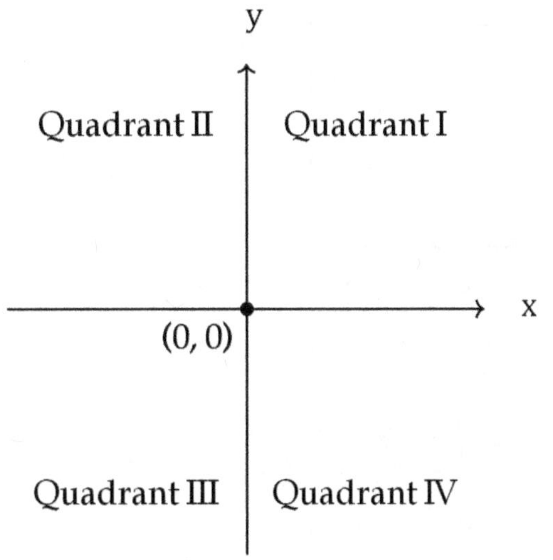

Figure 4.5: The Coordinate Plane showing the four quadrants and the origin.

Example 1:

Identifying Quadrants For the point (−2, 3): - x = −2 (which means it's to the left of the origin), - y = 3 (which means it's above the origin).

Thus, the point (−2, 3) is located in **Quadrant II**.

For the point (4, −5): - x = 4 (which means it's to the right of the origin), - y = −5 (which means it's below the origin).

Thus, (4, −5) is located in **Quadrant IV**.

4.2.6 Plotting Points

To plot a point, you locate its position by moving along the x-axis and then moving parallel to the y-axis. For example, to plot the point (3, −2), you: 1. Start at the origin, 2. Move 3 units to the right (because x = 3), 3. Then move 2 units down (because y = −2).

The point (3, −2) lies in **Quadrant IV**.

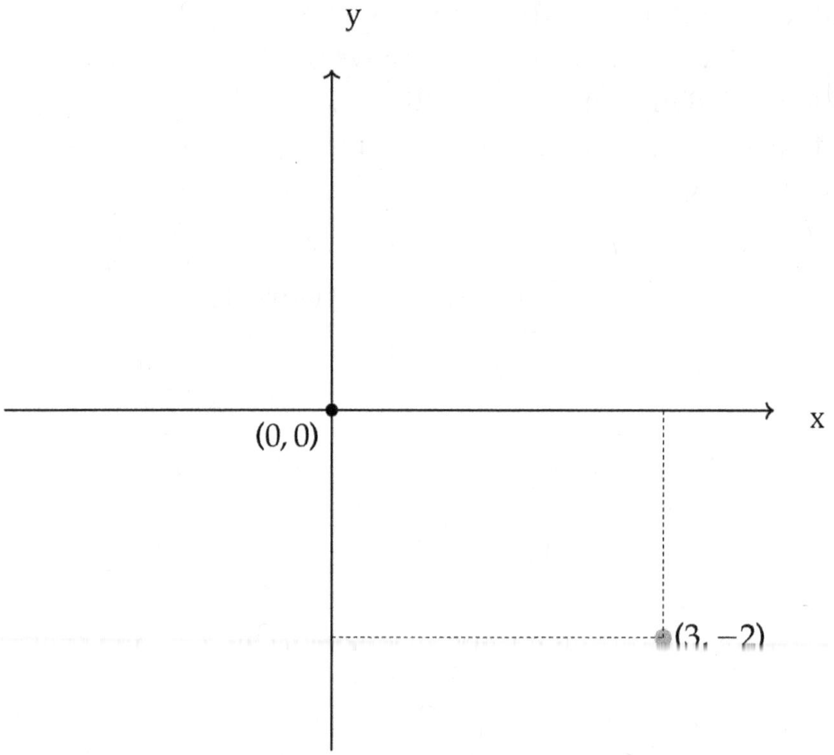

Figure 4.6: Plotting the point (3, −2) on the coordinate plane.

4.2.7 Calculating Slope and Distance Between Points

Understanding how to calculate the slope and the distance between points is critical in solving many GED geometry and algebra questions.

Calculating Slope

The slope of a line measures how steep a line is and is calculated as the ratio of the change in the y-coordinates (rise) to the change in the x-coordinates (run) between two points:

$$slope = m = \frac{y_2 - y_1}{x_2 - x_1}$$

Example 2: Calculating the Slope Between Two Points
Given two points, A (1, 2) and B (4, 6), find the slope of the line passing through them.

Using the slope formula:

$$m = \frac{6-2}{4-1} = \frac{4}{3}$$

So, the slope of the line through points A and B is $\frac{4}{3}$.

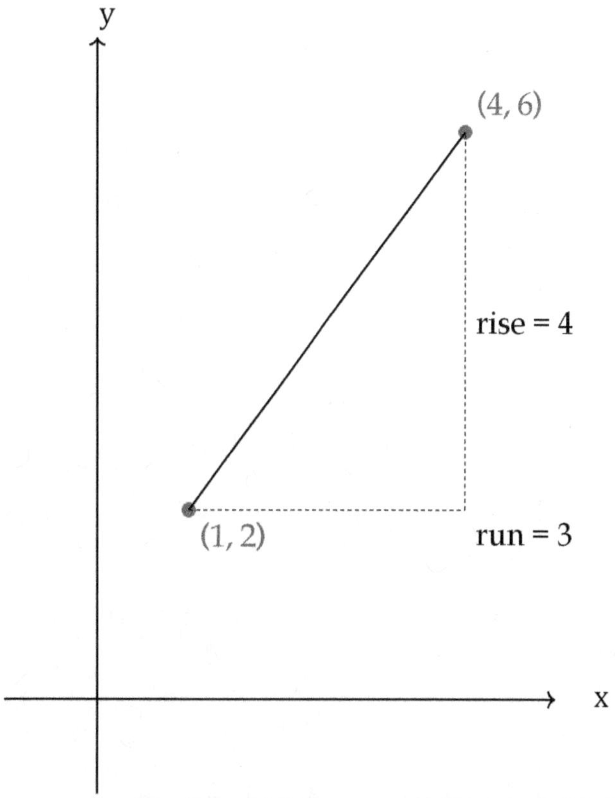

Figure 4.7: Calculating slope: Line between points (1, 2) and (4, 6) with rise and run.

Calculating the Distance Between Two Points

The distance d between two points (x_1, y_1) and (x_2, y_2) on a coordinate plane can be found using the distance formula, which is derived from the Pythagorean theorem:

$$d = \sqrt{(x_2 - x_1)^2 + (y_2 - y_1)^2}$$

Example 3: Finding the Distance Between Two Points
Using the same points A (1, 2) and B (4, 6) from the previous example, we can calculate the distance between them:

$$d = \sqrt{(4-1)^2 + (6-2)^2} = \sqrt{3^2 + 4^2} = \sqrt{9 + 16} = \sqrt{25} = 5$$

Thus, the distance between A (1, 2) and B (4, 6) is 5 units.

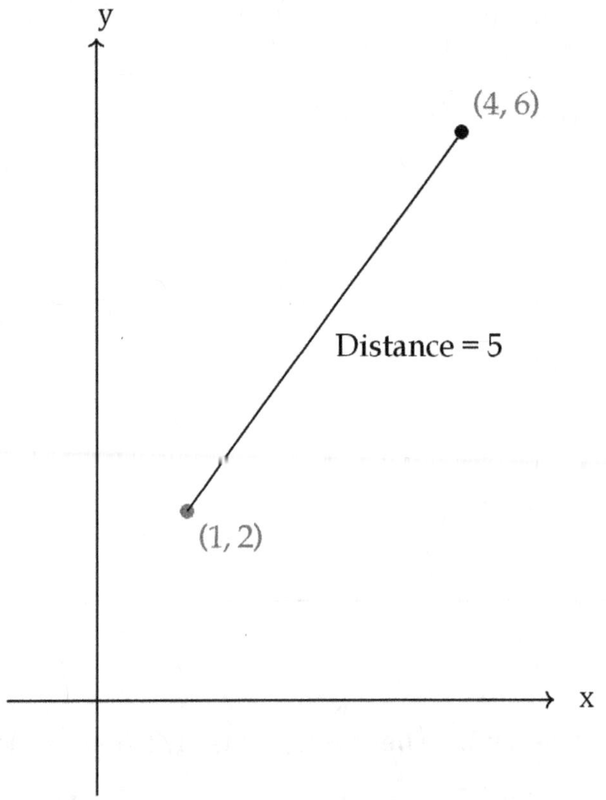

Figure 4.8: Finding the distance between two points (1, 2) and (4, 6).

4.2.8 Summary

In this section, we have covered the basics of the coordinate plane and learned how to plot points. We delved into calculating the slope and the distance between points, which are essential skills for understanding relationships between different points and lines in coordinate geometry. Mastering these concepts will improve your problem- solving abilities and prepare you for questions on the GED Math test involving graphs, lines, and spatial reasoning.

Exercises

Exercise 4.7.

Plot the following points on the coordinate plane and label them: A (3, 4), B (−2, 1), C (0, −3), and D (−4, 5). After plotting, determine in which quadrant each point lies.

Exercise 4.8.

Calculate the slope of the line passing through the points P (2, 7) and Q (−3, −1). Show your work using the slope formula and explain if the slope is positive, negative, or zero.

Exercise 4.9.

Given the points X (1, 2) and Y (4, 6), use the distance formula to calculate the distance between the two points. Express your answer in simplest form.

Answers with Explanations

Solution 4.7. Plotting Points and Quadrant Identification:

To plot the points, we can use a simple coordinate grid where the x-axis is horizontal and the y-axis is vertical. The coordinates indicate the position relative to the origin (0, 0). Figure 4.9 shows the points A (3, 4), B (−2, 1), C (0, −3), and D (−4, 5) plotted on the coordinate plane.

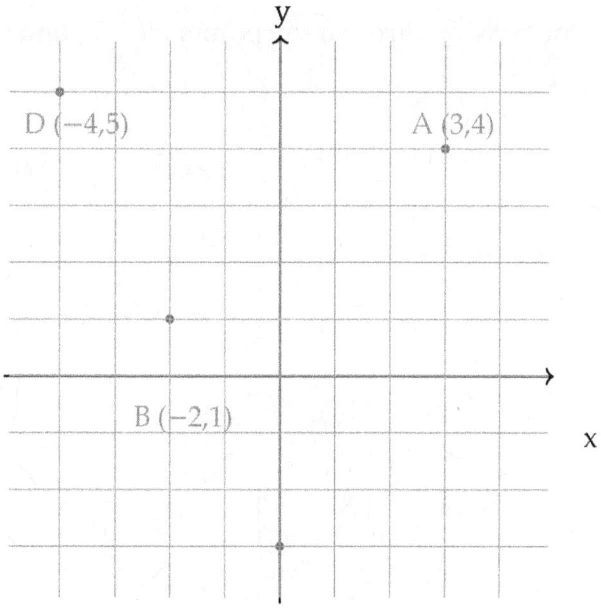

C (0,–3)

Figure 4.9: Points plotted on the coordinate plane

In terms of quadrant identification:

- Point A $(3, 4)$ is in the first quadrant (positive x, positive y).
- Point B $(-2, 1)$ is in the second quadrant (negative x, positive y).
- Point C $(0, -3)$ is on the negative y-axis.
- Point D $(-4, 5)$ is in the second quadrant (negative x, positive y).

Solution 4.8.

Finding the Slope of Line Between P $(2, 7)$ and Q $(-3, -1)$:

The formula to calculate the slope m between two points (x_1, y_1) and (x_2, y_2) is:

$$m = \frac{y_2 - y_1}{x_2 - x_1}$$

Substituting the points P $(2, 7)$ and Q $(-3, -1)$:

$$m = \frac{-1 - 7}{-3 - 2} = \frac{-8}{-5} = \frac{8}{5}$$

Therefore, the slope of the line is Since the slope is positive, this means the line rises as it moves from left to right on the coordinate plane.

Below is a plot of the line passing through the points P $(2, 7)$ and Q $(-3, -1)$.

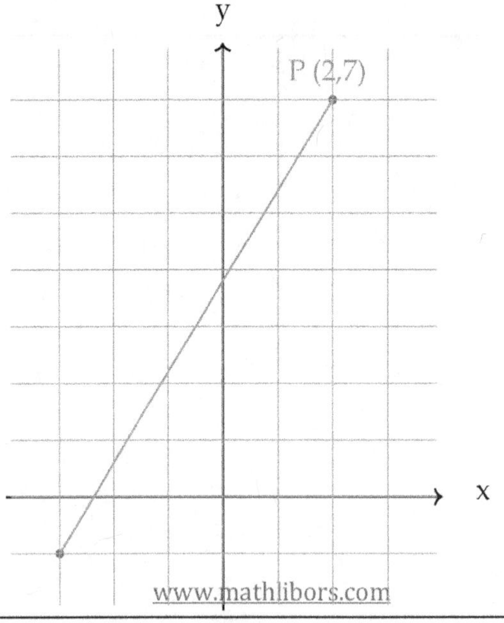

P (-3, -1)

Figure 4.10: Line passing through P (2, 7) and Q (−3, −1)

Solution 4.9.

Distance Between Points X $(1, 2)$ **and** Y $(4, 6)$:

The distance formula is given by:

$$d = \sqrt{(x_2 - x_1)^2 + (y_2 - y_1)^2}$$

Substituting the coordinates of points X $(1, 2)$ and Y $(4, 6)$:

$$d = \sqrt{(4 - 1)^2 + (6 - 2)^2} = \sqrt{3^2 + 4^2} = \sqrt{9 + 16} = \sqrt{25} = 5$$

Therefore, the distance between points X and Y is 5 units.

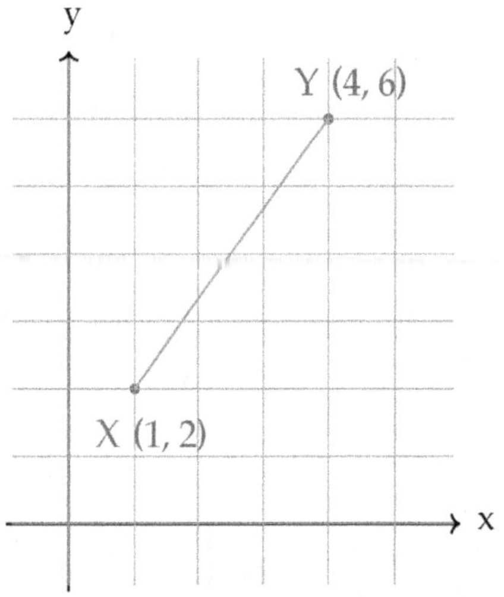

Figure 4.11: Line segment representing the distance between X $(1, 2)$ and Y $(4, 6)$

Section: Transformations

In this section, we will dive deep into the transformations of geometric figures in a plane. Transformations allow us to manipulate shapes in various ways while preserving certain properties, such as distance, angle measurements, and relative orientation. A transformation can be thought of as an operation that moves or changes a shape in some way, and it includes translations, reflections, rotations, dilations, and scaling.

Transformations are essential in understanding the properties of objects in both Euclidean and analytic geometry. They often connect algebraic concepts with geometric understanding, an important skill on the GED Math test.

Understanding Translations

A "translation" is a type of transformation that "slides" a figure from one location to another on the coordinate plane without changing its shape, size, or orientation. Every point of the shape moves the same distance in the same direction.

Consider a point P (x, y), and you are asked to translate this point by adding a units to the x-coordinate and b units to the y-coordinate. The translation can be described as:

$$T_a^b(P) = P'(x + a, y + b)$$

In this translation, the point P moves horizontally by a unit and vertically by b units.

Example 1:
Translating a Point Suppose we have a point A (2, 3), and we wish to translate it 4 units to the right and 2 units up. The new coordinates A′ after the translation are calculated as follows:

$$A' = (2 + 4, 3 + 2) = (6, 5)$$

Thus, A (2, 3) becomes A′ (6, 5) after the translation.

Reflections and Rotations

Reflections: A "reflection" is a transformation that produces a mirror image of a shape over a given line, called the "line of reflection". The figure is flipped over the line, and each point on the original figure moves perpendicularly across the line of reflection.

The most common lines of reflection in the coordinate plane include the x-axis, the y-axis, or lines like y = x.

- Reflection over the x-axis: A point P (x, y) is reflected over the x-axis, resulting in P′ (x, −y).
- Reflection over the y-axis: A point P(x, y) is reflected over the y-axis, resulting in P′ (−x, y).

Example 2:
Reflecting a Point Over the x-axis Let's reflect the point B (4, 5) over the x-axis. The new coordinates after reflection are calculated as follows:

$$B' = (4, -5)$$

Therefore, B (4, 5) reflects to B′ (4, −5).

Rotations: A "rotation" is a transformation that turns a figure around a fixed point, commonly called the "center of rotation", by a certain angle and in a specific direction (usually counterclockwise unless otherwise specified).

When discussing rotations, particularly on the coordinate plane, 90°, 180°, and 270°

rotations are common. Standard rotations around the origin (0, 0) follow these rules:

- 90° counterclockwise rotation: $P(x, y) \rightarrow P'(-y, x)$
- 180° counterclockwise rotation: $P(x, y) \rightarrow P'(-x, -y)$
- 270° counterclockwise rotation: $P(x, y) \rightarrow P'(y, -x)$

Example 3:

Rotating a Point 90° Counterclockwise Rotate the point C (3, 2) 90° counterclockwise around the origin. Applying the rule for 90° rotation gives:

$$C' = (-2, 3)$$

So, C (3, 2) becomes C′ (−2, 3) after a 90° rotation.

Dilations and Scaling

"Dilations" are transformations that change the size of a figure without altering its shape. In a dilation, every point on the figure moves closer to or farther away from a fixed point, known as the "center of dilation", by a factor called the "scale factor".

- When the scale factor is greater than 1, the figure enlarges.
- When the scale factor is between 0 and 1, the figure shrinks.

The new coordinates of a point P (x, y) after a dilation with scale factor k, centered at the origin, are given by:

$$P' = (kx, ky)$$

Example 4:

Dilating a Point If we dilate the point D (2, 4) by a scale factor of k = 1.5, the new coordinates are:

$$D' = (1.5 \times 2, 1.5 \times 4) = (3, 6)$$

Thus, the point D (2, 4) dilates to D′ (3, 6).

Figure 1: Dilation of a Triangle Below is an illustration showing the dilation of a triangle by a scale factor of 2 centered at the origin.

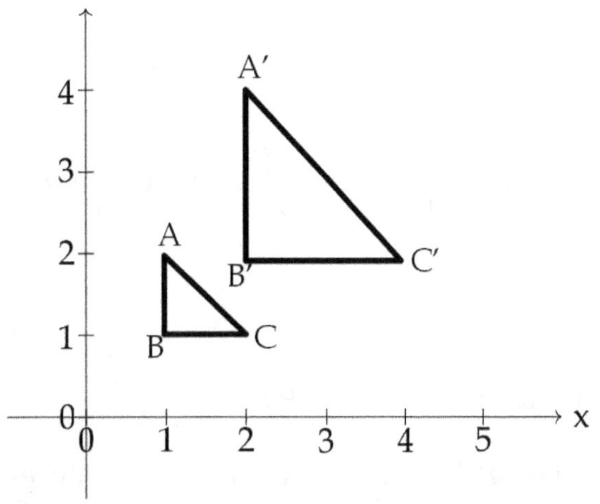

As illustrated, the original triangle ABC is dilated by a factor of 2 to create the larger A′B′C′.

Conclusion

Understanding how to manipulate figures through transformations like translations, reflections, rotations, and dilations is essential for solving many geometric problems on the GED Math test. With practice, you will become proficient at recognizing and applying these transformations in various contexts.

Exercises

Exercise 4.10.

1. A point (x, y) is translated 3 units to the right and 5 units up. If the original coordinates of the point are (2, 4), what are the new coordinates after the translation?

2. A triangle with vertices at (2, 3), (4, 6), and (5, 1) is reflected across the x-axis.

What are the coordinates of the new vertices after the reflection?

3. A square has vertices at (1, 1), (1, 4), (4, 4), and (4, 1). Scale the square by a factor of 2 centered at the origin. What are the coordinates of the new vertices after the scaling?

Exercise 4.11.

A polygon undergoes a counterclockwise rotation of 90°about the origin. The point A (3, 2) is one of the vertices of the polygon. What are the new coordinates of A after the transformation?

Exercise 4.12.

Given the triangle with vertices A (1, 2), B (3, 4), and C (5, 1), perform a dilation with a scale factor of 0.5 centered at the origin. Find the new coordinates of A, B, and C after the dilation.

Answers with Explanations

Solution 4.10. Exercise 1:

1. The point's original coordinates are (2, 4). If we translate the point 3 units to the right, we add 3 to the x-coordinate, resulting in $2 + 3 = 5$. Similarly, translating the point 5 units up means adding 5 to the y-coordinate, giving $4 + 5 = 9$. Thus, the new coordinates are (5, 9).

2. The new coordinates after reflecting the triangle with vertices (2, 3), (4, 6), and (5, 1) across the x-axis, result in (2, −3), (4, −6), and (5, −1), respectively. The y-coordinates are negated during reflection across the x-axis.

3. The vertices of the square are scaled by a factor of 2, meaning we multiply each coordinate by 2, centered at the origin (0, 0).

$$(1,1) \rightarrow (2,2), \quad (1,4) \rightarrow (2,8), \quad (4,4) \rightarrow (8,8), \quad (4,1) \rightarrow (8,2)$$

So, the new vertices are (2, 2), (2, 8), (8, 8), and (8, 2).

Solution 4.11. Exercise 2:

When we rotate the point A (3, 2) counterclockwise by 90° about the origin, the rule for such rotations is:

$$(x', y') = (-y, x)$$

Hence, the rotation of A (3, 2) results in the new coordinates:

$$A'(-2, 3)$$

Therefore, after the 90° counterclockwise rotation, the new coordinates of A are (−2, 3).

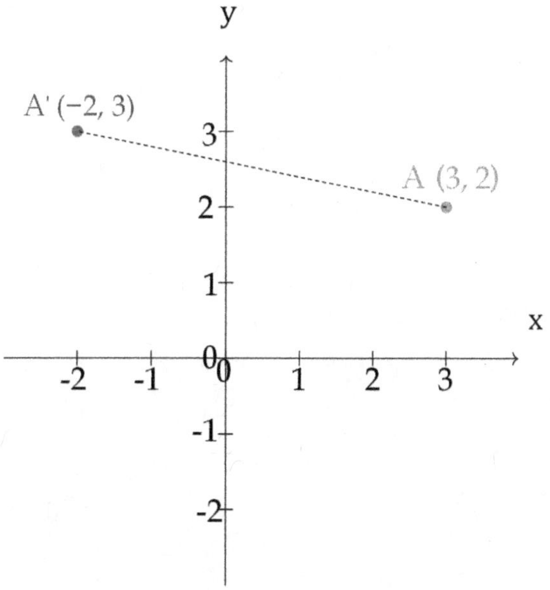

Solution 4.12. Exercise 3:

The dilation of the triangle with vertices A (1, 2), B (3, 4), and C (5, 1) by a factor of 0.5 centered at the origin involves multiplying each coordinate by 0.5:

$$A(1,2) \rightarrow \left(\frac{1}{2}, 1\right) = (0.5,1)$$

$$B(3,4) \rightarrow \left(\frac{3}{2}, 2\right) = (1.5,2)$$

$$C(5,1) \rightarrow \left(\frac{5}{2}, \frac{1}{2}\right) = (2.5,0.5)$$

Hence, the new coordinates after the dilation are:

A′ (0.5, 1), B′ (1.5, 2), C′ (2.5, 0.5)

y

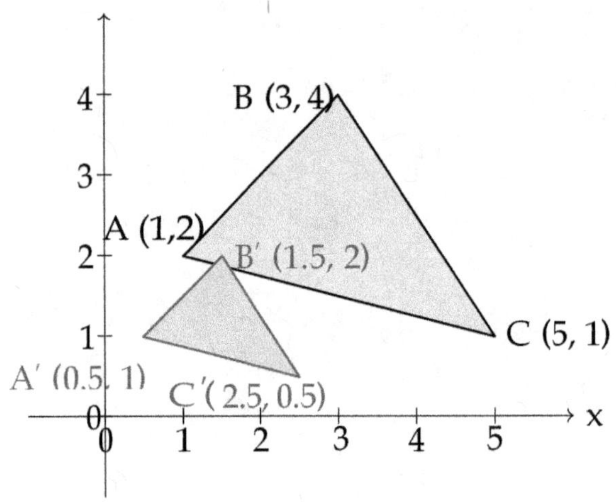
B (3, 4)
A (1,2)
B′ (1.5, 2)
C (5, 1)
A′ (0.5, 1)
C′(2.5, 0.5)
x
0
1
2
3
4
5

CHAPTER # 5

Data Analysis, Statistics, and Probability

Math topics that you'll learn in this chapter:

- Understanding Data
- Types of Data (Quantitative vs. Qualitative)
- Reading Tables and Charts
- Creating Graphs and Plots
- Measures of Central Tendency
- Calculating Mean
- Calculating Median
- Understanding Mode
- Comparing Mean, Median, Mode
- Probability
- Fundamentals of Probability
- Probability of Single Events
- Compound Probability
- Statistical Questions
- Understand Range and Variability
- Interpreting Data Distributions
- Recognizing Misleading Graphs and Data

5.1 Understanding Data

Understanding how data is collected, categorized, and presented is an essential skill, not just for the GED Math Test but also for various real-life situations. In this section, we will cover the types of data, how to read tables and charts, and how to create graphs and plots. Mastery of this topic will allow you to make sense of collected data, interpret it effectively, and make informed decisions based on statistical information.

5.1.1 Types of Data (Quantitative vs. Qualitative)

Data can broadly be classified into two main types: *quantitative* and *qualitative* data. Understanding the distinction between these two types is crucial for handling data correctly and using the appropriate methods to analyze it.

Quantitative Data

Quantitative data refers to any data that is numerical. This type of data answers questions like "How many?" or "How much?". It can be further divided into:

1. Discrete Data: Values that can be counted, such as the number of students in a classroom or the number of cars in a parking lot.
2. Continuous Data: Values that can take any value within a range, such as height, weight, or temperature.

Example:

- The number of apples a store sells each day (discrete).
- The temperature readings over a day from a thermometer (continuous).

Qualitative Data

Qualitative data, also known as categorical data, represents characteristics or qualities that cannot be quantified numerically but can be categorized. This type of data answers questions like "What type?" or "What category?".

1. Nominal Data: Categories with no implied order. For example, eye color (blue, green, brown) or types of cars (sedan, SUV, truck).
2. Ordinal Data: Categories that have a meaningful order or ranking. For example, customer satisfaction ratings (poor, average, good, excellent) or education levels (high school, college, graduate school).

Example:

- Types of fruits sold at a market (apple, banana, orange) (nominal).

- Rankings in a race (1st, 2nd, 3rd place) (ordinal).

5.1.2 Reading Tables and Charts

When data is presented in tables or charts, the ability to extract and interpret the necessary information is key. This skill is tested on the GED Math Test and is an important life skill, as data summaries are frequently used in many fields, from business reports to scientific studies.

Understanding Tables

A table organizes data into rows and columns, making it easier to see patterns, compare values, and retrieve specific information. Let's consider the following example:

Product	**Units Sold**	**Revenue ($)**
Apples	500	1500
Bananas	700	2100
Oranges	550	1650

Table 5.1: Sales Data for Fruits in 2022

From this table, we can easily deduce the following:

- Bananas were the highest-selling product, with 700 units sold.
- Revenues from apples amounted to $1500, while oranges produced $1650.

Reading Charts

Charts provide a visual representation of data. Common types of charts include bar charts, pie charts, and line graphs.

Bar Chart Example

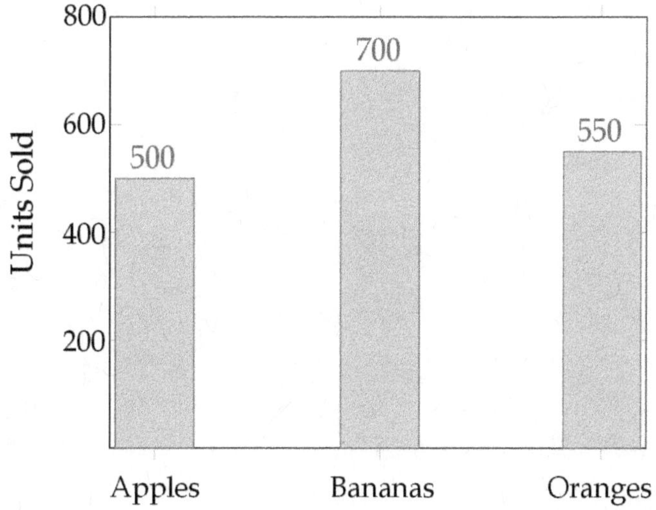

Products

Figure 5.1: Bar Chart Representing Fruit Sales

In this bar chart, it visually becomes clear that Bananas had the highest sales, followed by Oranges and then Apples. Bar charts are extremely useful when comparing discrete categories.

Interpreting Pie Charts

Pie charts are useful for showing the proportions of a whole. Take the following example, showing market share:

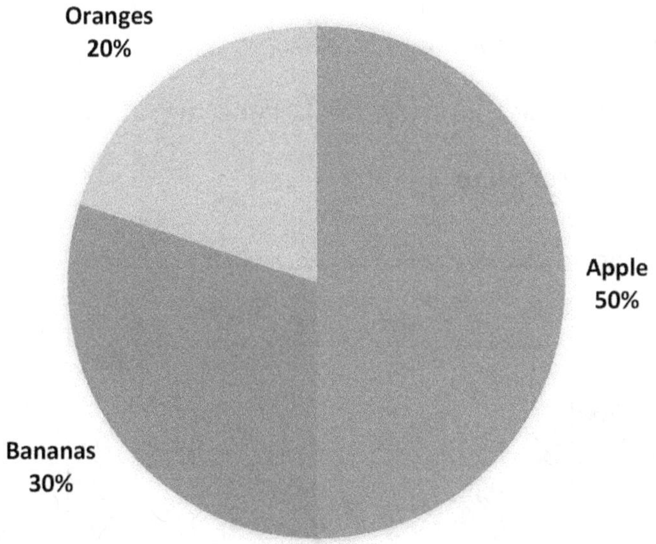

Figure 5.2: Market Share of Fruit Sales

Here, we see that apples take up 50% of the market share, bananas make up 30%, and oranges 20%, indicated clearly by the size of each "slice" of the pie chart.

5.1.3 Creating Graphs and Plots

Being able to create graphs and plots from raw data is another fundamental skill. Below, we will walk you through how to create basic bar, line, and scatter plots.

Bar Graph

A bar graph can be created when you have categorical data. To create a bar graph, follow these steps:

- Label the x-axis with the categories (e.g., apples, bananas, oranges).
- Label the y-axis with the quantity being measured (e.g., number of units sold).
- Draw bars for each category proportional to their value.

Line Graph

Line graphs represent how data changes over time. Let's say we have the following data for monthly sales:

Months = {January, February, March, April} Units Sold = {150, 200, 180, 220}

To create a line graph:

- Label the x-axis with the months.
- Label the y-axis with the number of units sold.
- Plot points at each value (e.g., (January, 150), (February, 200)).
- Connect the points to form a line.

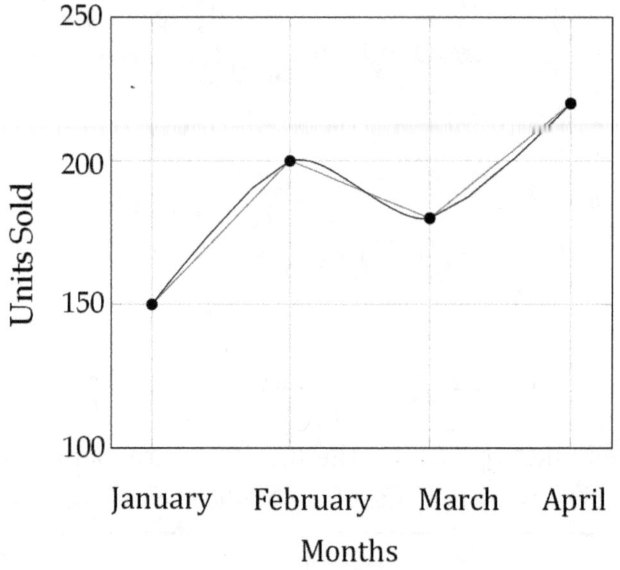

Figure 5.3: Line Graph Displaying Monthly Sales

Scatter Plot

Scatter plots are used to represent the relationship between two sets of data. Suppose we have the following data showing the time spent studying vs. test scores:

Hours Studied = {1, 2, 3, 4, 5}

Test Score = {60, 65, 70, 75, 80}

To create a scatter plot:

- Label the x-axis as hours studied.
- Label the y-axis as the test score.

- Plot each data point (e.g., (1, 60), (2, 65)).

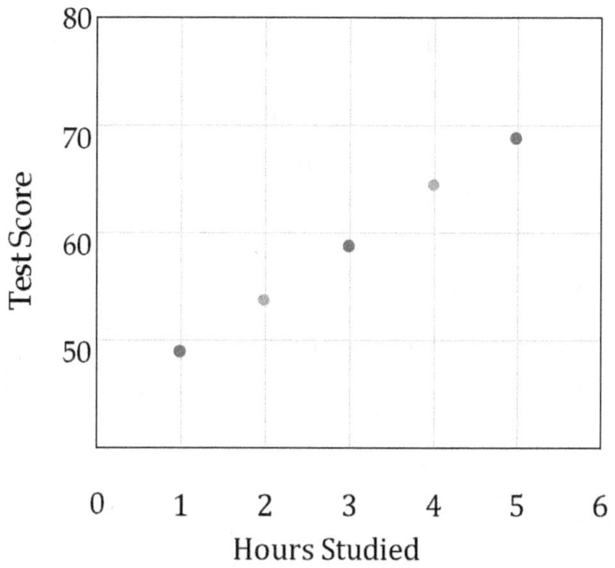

Figure 5.4: Scatter Plot Illustrating Hours Studied vs. Test Score

Here, each point represents a separate test score with respect to the hours studied. You can observe a positive trend, meaning that as the time spent studying increases, the test score improves.

With these visual tools in your toolkit, you now have the ability to not only read and understand data better but also to create your own charts, tables, and graphs for summarizing information. This skill will be invaluable in both the GED Math Test and your everyday decision-making.

Exercises

Exercise 5.1.

For the following dataset, determine whether each variable is quantitative or qualitative:

Variable	Description
Ages	The age of a person (in years)
Gender	Gender identity (e.g., male, female, non-binary)
Income	Monthly Income (in dollars)
Grade Level	Current academic grade (e.g., 9th, 10th, 11th)

Exercise 5.2.

A survey asked respondents to describe their favorite type of music. The responses were grouped into the following categories: Rock, Pop, Jazz, Classical, Country. Draw a bar chart representing the data below:

Music Type	Number of Respondents
Rock	45
Pop	30
Jazz	15
Classical	10
Country	20

Use the data provided to create a labeled bar graph.

Exercise 5.3.

Create a scatter plot for the data points provided below. Clearly identify the axes and label each point. After plotting the points, describe if there is any pattern or relationship between the variables.

X Value	Y Value
1	2
2	3
3	5
4	7
5	8

Answers with Explanations

Solution 5.1.

1. **Age**: Quantitative, since age is a numerical value and can be measured.
2. **Gender**: Qualitative, since gender is a category and does not represent a numerical value.
3. **Income**: Quantitative, as income is measurable in specific units (dollars).

4. Grade Level: Qualitative, since grade levels represent classifications but are not absolute numerical measurements.

Solution 5.2.

The bar chart is created as follows:

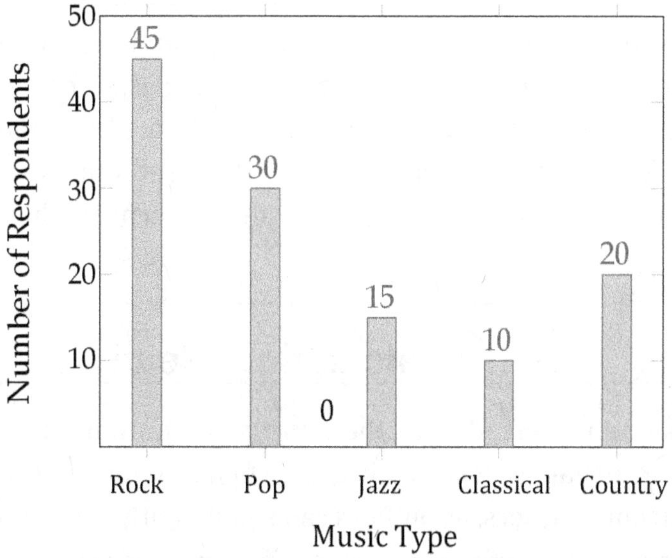

The peak popularity is in the "Rock" category, with the lowest number of respondents favoring "Classical."

Solution 5.3.

The scatter plot is interpreted as follows:

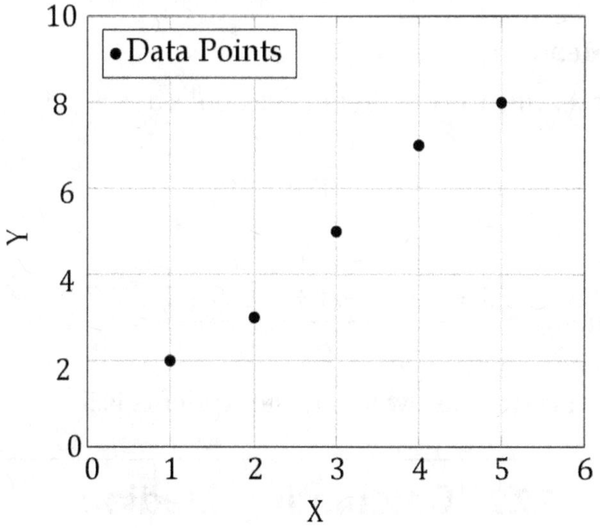

The scatter plot suggests a positive linear relationship. As the X-value increases, the Y-value also tends to increase, indicating a potential correlation.

5.2 Measures of Central Tendency

Measures of central tendency are critical concepts in understanding data distribution. These measures summarize a dataset by identifying a single value that represents a center point or typical value. The most commonly used measures are the *mean*, *median*, and *mode*. Each provides different insights about the data and is useful in various contexts. In this section, we will explore how to calculate and interpret these three measures.

5.2.1 Calculating Mean

The *mean*, which is often referred to as the *average*, is the sum of all values in a dataset divided by the number of values. It is widely used when all data points are relatively similar and there are no extreme outliers, as outliers can significantly affect the mean.
The formula for calculating the mean of a dataset is expressed as:

$$Mean = \frac{\sum_{i=1}^{n} x_i}{n}$$

Where:

- means the sum of all values
- represents each individual value in the dataset,
- n is the number of data points.

Example: Calculating Mean

Consider a dataset representing the number of textbooks a student owns:

$$3, 5, 2, 8, 4, 6, 7$$

To find the mean:

$$mean = \frac{3 + 5 + 2 + 8 + 4 + 6 + 7}{7} = \frac{37}{5} = 5$$

Thus, the mean number of textbooks owned by the students is 5.

5.2.2 Calculating Median

The *median* is the middle value in a dataset when the values are arranged in numerical order. If the dataset has an odd number of values, the median is simply the middle value. If the dataset contains an even number of values, the median is the average of the two middle numbers.

To compute the median:

1. Sort the data values in ascending order.
2. If the number of values (n) is odd, the median is the value at the position $\frac{n+1}{2}$.
3. If n is even, the Revenue (median is the average of the values at the positions $\frac{n}{2}$ *and* $\frac{n}{2}+1$.

Example 1: Odd Number of Data Points (Median)
Let's consider the following dataset, representing the number of hours students spent studying for a test:

$$4, 1, 7, 3, 9$$

First, we sort the values in ascending order:

$$1, 3, 4, 7, 9$$

Since we have 5 points (odd number), the median is simply the middle value, which is 4.

Example 2: Even Number of Data Points (Median)
Now, consider another dataset:

8, 12, 3, 5

First, we arrange the values in order:

3, 5, 8, 12

Since there are 4 points (even number), the median is the average of the two middle numbers (5 and 8):

$$Median = \frac{5+8}{2} = 6.5$$

Thus, the median is 6.5.

5.2.3 Understanding Mode

The *mode* is the value that appears most frequently in a dataset. A dataset can have one mode (unimodal), more than one mode (bimodal or multimodal), or no mode at all if no value repeats.

Example 1: Unimodal (Single Mode)
Consider the following data representing quiz scores:

$$10, 7, 7, 9, 10, 10, 6, 8$$

Here, the mode is 10 because it appears three times, which is more than any other value.

Example 2: Bimodal (Two Modes)
Now, consider another dataset:

$$3, 5, 7, 5, 8, 3, 9$$

In this case, both 3 and 5 appear twice, meaning this dataset is *bimodal*, with both 3 and 5 as the modes.

Example 3: No Mode
In the dataset:

$$2, 4, 6, 8, 10$$

Each value appears only once. Thus, this dataset has no mode.

5.2.4 Comparing Mean, Median, and Mode

Each measure of central tendency provides valuable information, but they are most informative when used in conjunction or depending on the distribution of the data.

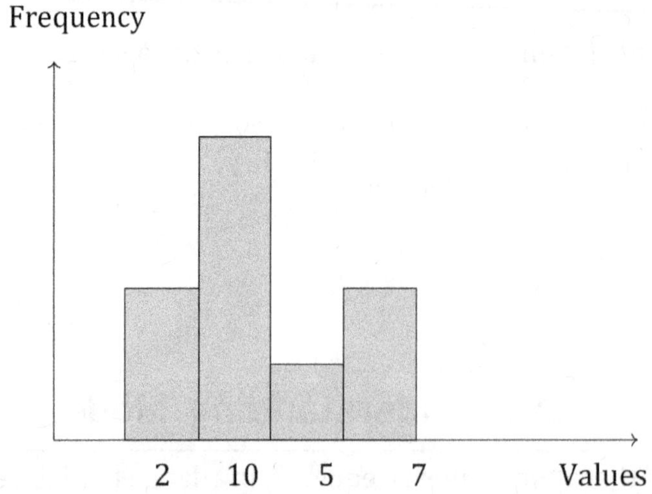

Figure 5.5: A simple unimodal dataset: The mode is 10, as it occurs most frequently.

Extreme values (*outliers*) can affect the mean significantly, whereas the median is more robust to outliers because it depends only on the middle value(s), making it a better measure in skewed distributions. The mode is useful when identifying the most frequent or popular value, which can be particularly insightful for categorical data.

Measure of Central Tendency	Affected by Outliers	Best for Use With	Does it Always Exist
Mean	Yes	Quantitative Data	Yes
Median	No	Skewed Data	Yes
Mode	No	Categorical Data, Repeated Values	No

Table 5.2: Comparison of Measures of Central Tendency

Summary

Understanding measures of central tendency like the mean, median, and mode is fundamental for analyzing and interpreting data. Each has its strengths and weaknesses depending on the context and characteristics of the dataset. By applying each appropriately, we can capture valuable insights from different types of data, whether it is numerical, categorical, or highly skewed. Moreover, these measures form the basis for further statistical analysis and decision-making in real-world scenarios.

Exercises

Exercise 5.4.

1. Calculate the mean of the following data set: $\{8, 12, 15, 22, 16, 19, 29\}$. Show your calculations step by step.

2. Given the following set of numbers: $\{34, 22, 17, 17, 35, 40, 35, 17, 40\}$, determine the median. If necessary, explain how to handle repeated numbers.

3. Consider the following set of scores: $\{65, 70, 80, 75, 85, 90, 80, 85, 80\}$. Identify the mode and describe why the mode is significant.

Exercise 5.5.

A teacher recorded the following test scores from a class:

$$\mathbf{52, 74, 89, 45, 67, 93, 88, 74, 88, 95}$$

1. **Calculate the mean score.**
2. **Determine the median score.**
3. **Identify the mode of the above scores, if it exists.**

Exercise 5.6.

You are given the following number of hours students spent studying for a math test:

$$\{3, 3, 4, 5, 6, 7, 8, 8, 9, 12, 12\}$$

1. **Compute the mean of the data.**
2. **Find the median.**
3. **Identify the mode of the data, if it exists.**

Exercise 5.7.

Use the following dataset representing ages of individuals in a group:

$$25, 30, 31, 35, 35, 28, 26, 35, 29, 30, 35$$

1. **Calculate the mean age of the group.**
2. **Determine the mode, if it exists.**
3. **Calculate the median age of the group.**

Answers with Explanations

Solution 5.4. Exercise 1:

1. To calculate the mean:

$$Mean = \frac{8 + 12 + 15 + 22 + 16 + 19 + 29}{7} = \frac{121}{7} \approx 17.29$$

So, the mean is approximately 17.29.

2. The data set is: $\{34, 22, 17, 17, 35, 40, 35, 17, 40\}$. Arranging in order:

$$17, 17, 17, 22, 34, 35, 35, 40, 40$$

The middle value is 34, so the median is 34.

3. The mode is the number that appears most frequently. Here, the mode is 80, as it appears 3 times.

Solution 5.5. Exercise 2:

1. To calculate the mean:

$$Mean = \frac{52 + 74 + 89 + 45 + 67 + 93 + 88 + 74 + 88 + 95}{10} = \frac{765}{10} = 76.5$$

So, the mean score is 76.5.

2. Arranging the scores:

$$45, 52, 67, 74, 74, 88, 88, 89, 93, 95$$

The median score is the average of the 5th and 6th values:

$$\frac{74+88}{2}=\frac{162}{2}=81$$

3. The modes are 74 and 88, as both appear twice.

Solution 5.6. Exercise 3:

1. To calculate the mean:

$$Mean=\frac{3+3+4+5+6+7+8+8+9+12+12}{11}=\frac{77}{11}\approx 7$$

2. Arrange the data in order:

$$3, 3, 4, 5, 6, 7, 8, 8, 9, 12, 12$$

The median is the middle value which is 7.

3. The mode is 3, as it appears more frequently than the other numbers.

Solution 5.7. Exercise 4:

1. Mean:

$$Mean=\frac{25+30+31+35+35+28+26+35+29+30+35}{11}=\frac{339}{11}\approx 30.82$$

2. The mode is 35, as it appears 4 times in the dataset.

3. Median:

$$25, 26, 28, 29, 30, 30, 31, 35, 35, 35, 35$$

The middle value is 30, so the median is 30.

In this LaTeX code, we have written several exercises related to measures of central tendency (mean, median, mode), followed by the solutions that provide explanations along with calculations and necessary reasoning. In case diagrams or tables are required for explanation, 'tikz' and LaTeX table environments can be used accordingly.

5.3 Probability

Probability refers to the likelihood of an event occurring. When we're uncertain about the outcome of an event, we can express the chances of different results in terms of probability.

5.3.1 Fundamentals of Probability

Probability is a measure that quantifies the likelihood of a particular event happening. Events can either be certain, impossible, or somewhere in between. The probability of an event E happening is represented as P(E), and its value is always between 0 and 1:

- If P(E) = 0, the event will definitely *not* happen. - If P(E) = 1, the event *will* definitely happen. - If 0 < P(E) < 1, there is a chance that the event may or may not happen.

The formula for calculating the probability of an event E is given by:

$$P(E) = \frac{Number\ of\ favorable\ outcomes}{Total\ number\ of\ outcomes}$$

Let's go over a classic example involving rolling a die.

Example: Rolling a Die

Suppose we're rolling a fair six-sided die. The total number of possible outcomes when rolling one die is 6, as the die can land on any of the six faces.

1. What is the probability of rolling a 3?

The event E1 of rolling a 3 has only one favorable outcome (i.e., when the die shows the number 3). Therefore,

$$P(E_1) = \frac{1}{6}$$

2. What is the probability of rolling an odd number? The event E2 of rolling an odd number has three favorable outcomes (i.e., 1, 3, and 5). Therefore

$$P(E_2) = \frac{3}{6} = \frac{1}{2}$$

We can also represent each outcome of rolling a die visually. In Figure 5.6, the set of possible outcomes is shown as a sample space.

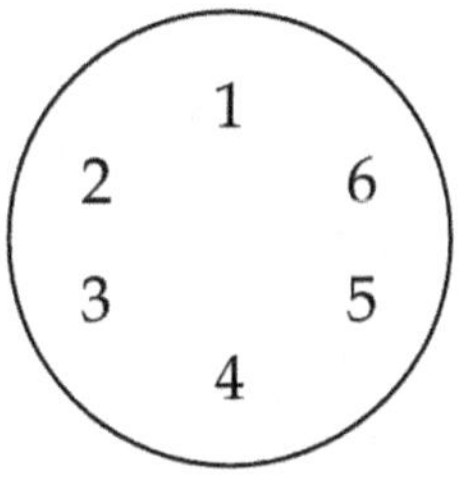

Figure 5.6: The Sample Space S for Rolling a Six-Sided Die

5.3.2 Probability of Single Events

In real-life scenarios, we often want to compute the probability of simple, single events. These events arise when only one outcome is of interest. Let's explore this concept through more examples.

Example: Coin Toss

Suppose we toss a fair coin. The set of possible outcomes is:

S = {Heads, Tails}

The number of outcomes in the sample space S is 2. Let's calculate the probability of specific events.

1. What is the probability of getting heads?

The event H representing heads has one favorable outcome ('Heads'). Therefore,

$$P(H) = \frac{1}{2}$$

2. What is the probability of getting tails?

The event T representing tails also has one favorable outcome ('Tails'). Thus, the probability of getting tails is:

$$P(T) = \frac{1}{2}$$

Notice that in both cases, the sum of the probabilities is:

$$P(H) + P(T) = \frac{1}{2} + \frac{1}{2} = 1$$

This illustrates a key property of probability: the sum of probabilities for all possible outcomes in a sample space must equal 1.

Example: Drawing a Card from a Deck of Cards

A standard deck of cards has 52 cards. Suppose a card is drawn at random. The number of outcomes in the sample space S is 52.

1. What is the probability of drawing an Ace?

In a deck of cards, there are four Aces (one for each suit: Hearts, Diamonds, Spades, and Clubs). The event A representing "drawing an Ace" has four favorable outcomes. Therefore, the probability is:

$$P(A) = \frac{4}{52} = \frac{1}{13}$$

2. What is the probability of drawing a red card?

There are 26 red cards in a deck (13 Hearts and 13 Diamonds). So, the probability of drawing a red card R is:

$$P(R) = \frac{26}{52} = \frac{1}{2}$$

5.3.3 Compound Probability

In many situations, we are interested in finding the probability of multiple events occurring together. This type of event is referred to as a *compound event.* Compound probability can be found by applying either the Addition Rule or the Multiplication Rule, depending on the types of events we are combining.

Addition Rule for Probability

The Addition Rule is used when we want to find the probability of either of two events happening. The rule varies depending on whether the events are mutually exclusive (i.e., they cannot occur at the same time) or not.

- **Mutually Exclusive Events:**

$$P\ (A \text{ or } B) = P(A) + P(B)$$

- **Non-Mutually Exclusive Events:** If events overlap (i.e., they can happen at the same time), we must subtract the probability of them both occurring:

$$P\ (A \text{ or } B) = P(A) + P(B) - P\ (A \text{ and } B)$$

Example: Mutually Exclusive Events

If you roll a die, the events "rolling a 3" and "rolling a 5" are mutually exclusive because the die can only show one face at a time. The probability of rolling either a 3 or a 5 is:

$$P(3\ or\ 5) = P(3) + P(5) = \frac{1}{6} + \frac{1}{6} = \frac{2}{6} = \frac{1}{3}$$

Example: Non-Mutually Exclusive Events

Now, consider the events "drawing a red card" and "drawing a 10" from a deck of cards. There are 26 red cards and 4 tens (one from each suit). Notice that two of these tens are red (10 of Hearts and 10 of Diamonds). Therefore, these events overlap, and we use the general Addition Rule:

$$\text{P (Red or 10)} = \text{P(Red)} + \text{P (10)} - \text{P (Red and 10)}$$

Substitute the known values:

$$P(Red\ or\ 10) = \frac{26}{52} + \frac{4}{52} - \frac{2}{52} = \frac{28}{52} = \frac{7}{13}$$

Multiplication Rule for Probability

For compound events where we are looking for the probability that two events A and B both occur, we use the Multiplication Rule:

- **Independent Events (events that do not affect each other):**

$$\text{P (A and B)} = \text{P(A)} \times \text{P(B)}$$

- **Dependent Events (one event affects the other):**

$$\text{P (A and B)} = \text{P(A)} \times \text{P(B|A)}$$

where P(B|A) is the conditional probability of B occurring, given A has already occurred.

Example: Independent Events

Suppose we flip two coins. The events "getting heads on the first flip" and "getting heads on the second flip" are independent because the outcome of the first flip does not affect the second.

The probability of getting heads on both flips is:

$$P(H\ and\ H) = P(H) \times P(H) = \frac{1}{2} \times \frac{1}{2} = \frac{1}{4}$$

Example: Dependent Events

Now, let's consider drawing two cards from a deck without replacement. The event "drawing an Ace on the first draw" and the event "drawing an Ace on the second draw" are dependent because after the first card is drawn, the deck changes.

The probability of drawing an Ace on the first card is:

$$P(A_1) = \frac{4}{52}$$

If we draw an Ace on the first draw, there are 3 Aces left in a deck of 51 cards, so the probability of drawing an Ace on the second draw given that we drew one on the first is:

$$P(A_2|A_1) = \frac{3}{51}$$

The probability of drawing two Aces in a row is:

$$P(A_1 \ and \ A_2) = \frac{4}{52} \times \frac{3}{51} = \frac{12}{2652} = \frac{1}{221}$$

Exercises

Exercise 5.8.

A bag contains 5 red marbles, 3 blue marbles, and 2 green marbles. If one marble is drawn randomly from the bag, what is the probability that:

1. **The marble is red?**
2. **The marble is blue?**
3. **The marble is not green?**

Exercise 5.9.

A standard six-sided die is rolled. Calculate the probability of rolling:

1. **A number greater than 4.**
2. **An even number.**
3. **A 3 or a 5.**

Exercise 5.10.

In a deck of 52 cards, what is the probability of drawing:

1. **A heart?**
2. **A face card (Jack, Queen, or King)?**
3. **A red card or an Ace?**

Exercise 5.11.

A box contains 4 yellow balls, 5 black balls, and 6 white balls. If two balls are drawn successively without replacement, what is the probability that:

1. **Both balls are black?**
2. **The first ball is white and the second is yellow?**

Answers with Explanations

Solution 5.8.

1. The total number of marbles is $5+3+2=10$.

$$P(red) = \frac{5}{10} = \frac{1}{2}$$

2.

$$P(blue) = \frac{3}{10}$$

3.

$$P(not\ green) = 1 - P(green) = 1 - \frac{2}{10} = \frac{8}{10} = \frac{4}{5}$$

Solution 5.9.

The sample space for a die roll is $\{1, 2, 3, 4, 5, 6\}$, so there are 6 possible outcomes.

1. A number greater than 4 means rolling a 5 or 6, so:

$$P(number > 4) = \frac{2}{6} = \frac{1}{3}$$

2. The even numbers are 2, 4, 6, so:

$$P(even) = \frac{3}{6} = \frac{1}{2}$$

3. Rolling a 3 or a 5 gives:

$$P(3\ or\ 5) = \frac{1}{6} + \frac{1}{6} = \frac{2}{6} = \frac{1}{3}$$

Solution 5.10.

In a standard deck of cards, there are 13 cards for each suit (hearts, diamonds, clubs, spades), 12 face cards in total (since each suit has a Jack, Queen, and King), and 26 red cards (13 hearts and 13 diamonds).

1. Probability of drawing a heart:

$$P(heart) = \frac{13}{52} = \frac{1}{4}$$

2. Probability of drawing a face card:

$$P(face\ card) = \frac{12}{52} = \frac{3}{13}$$

3. Probability of drawing a red card or an Ace. There are 26 red cards and 4 Aces. We must subtract the 2 red Aces (since they're double-counted):

$$P(red\ or\ ace) = \frac{26}{52} + \frac{4}{52} - \frac{2}{52} = \frac{28}{52} = \frac{7}{13}$$

Solution 5.11.
Total balls in the box: 4 + 5 + 6 = 15.

1. The probability of drawing two black balls without replacement:

$$P(both\ black) = \frac{5}{14} \times \frac{4}{14} = \frac{20}{210} = \frac{2}{21}$$

2. The probability of first drawing a white ball, then a yellow ball:

$$P(white\ then\ yellow) = \frac{6}{15} \times \frac{4}{14} = \frac{24}{210} = \frac{4}{35}$$

5.4 Statistical Operations

In this section, we will dive into the final components of mastering data analysis: understanding statistical operations, including **range and variability**, **data distributions**, and the nuances of **misleading graphs and data**. The ability to interpret and apply these concepts will be critical not only to passing the GED Math Test but also to analyzing real-world data effectively. Let's explore each of these subsections in detail and provide ample examples to solidify your understanding.

5.4.1 Understanding Range and Variability

One of the main goals of statistics is to understand how data varies. Variability gives us information about how far apart the different data points are from each other. There are several measures of variability, but we begin with the most foundational ones: the **range** and the more comprehensive **interquartile range (IQR)**.

Range

The **range** is the simplest way to measure variability. It calculates the difference between the largest and the smallest values in a dataset. Mathematically, the range can be represented as:

$$\text{Range} = \text{Max Value} - \text{Min Value}$$

For example, suppose we have the exam scores of 6 students as follows:

$$\{72, 85, 91, 78, 95, 82\}$$

The maximum score is 95, and the minimum score is 72. Therefore, the range is:

$$95 - 72 = 23$$

Thus, the range of this dataset is 23. While the range provides a basic idea of data spread, it doesn't account for how values are distributed within the dataset, which leads us to other measures of variability such as interquartile range and standard deviation.

Interquartile Range (IQR)

The **interquartile range (IQR)** measures the spread of the middle 50% of the data. This measure is more robust than the range as it is not influenced by extreme values (outliers). To compute the IQR, we first need to divide the data into quartiles:

1. **Q1 (Lower Quartile):** The median of the lower half of the data.
2. **Q3 (Upper Quartile):** The median of the upper half of the data.

The IQR is then computed as:

$$IQR = Q3 - Q1$$

Let's continue with our example data:

$$\{72, 78, 82, 85, 91, 95\}$$

The dataset is already ordered. The median of the dataset is the average of the middle two values:

$$\frac{82 + 85}{2} = 83.5$$

To find Q1, we take the median of the lower half: {72, 78, 82}. The median is 78. For Q3, we take the median of the upper half: {85, 91, 95}. The median is 91. Therefore, the IQR is:

$$91 - 78 = 13$$

The IQR tells us that the middle 50% of the data is spread over a range of 13 points.

General Conclusion: While the range can give us basic boundaries for the spread of data, the IQR is much more useful in identifying the spread of the "central" portion of the data.

5.1.2 Interpreting Data Distributions

Understanding how data is distributed is crucial for recognizing patterns and making data-driven decisions. A **distribution** refers to the way in which the values of the dataset are spread across possible values.

Normal Distribution

One of the most common data distributions is the **normal distribution**. It is symmetric around the mean and has a bell-shaped curve, as seen in Figure 5.7. In a normal distribution:

- About 68% of the data falls within one standard deviation of the mean. - About 95% of the data falls within two standard deviations of the mean. - About 99.7% of the data falls within three standard deviations of the mean.

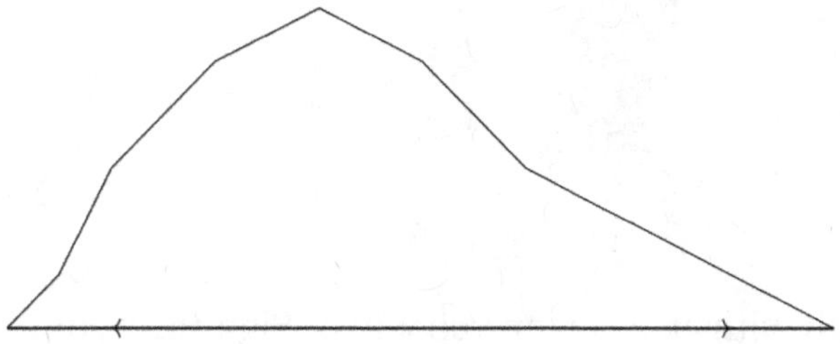

Mean
Normal Distribution

Figure 5.7: Symmetric normal distribution curve.

Skewed Distributions

Sometimes, data is not perfectly symmetrical. A dataset may be **left-skewed** (more data on the right) or **right-skewed** (more data on the left). Figures 5.8 and 5.9 illustrate these concepts.

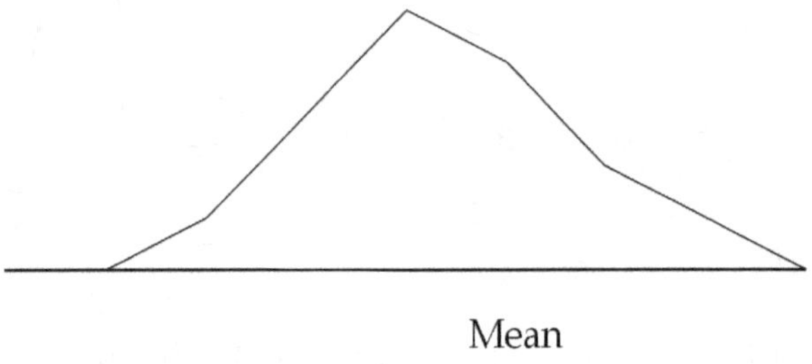

Figure 5.8: Left-skewed distribution.

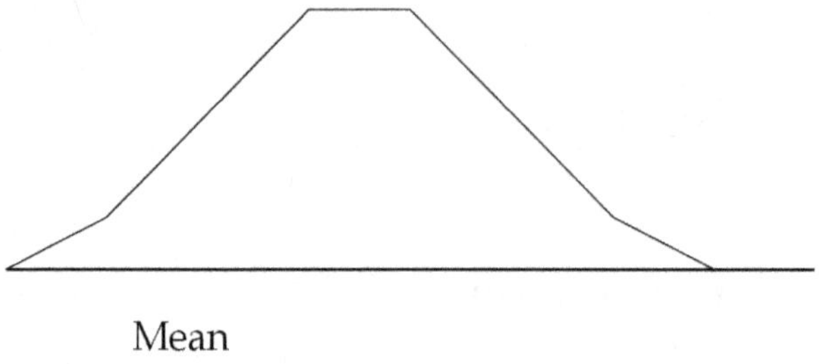

Figure 5.9: Right-skewed distribution.

5.1.3 Recognizing Misleading Graphs and Data

Graphs and charts are essential tools for visualizing data. However, sometimes graphs can be misleading, either intentionally or accidentally. Common problems include improper scaling, distorted axes, or omitting parts of the graph entirely.

Examples of Misleading Scales

Consider a bar graph where the y-axis does not start at zero (Figure 5.10). This practice can exaggerate minor differences between data points.

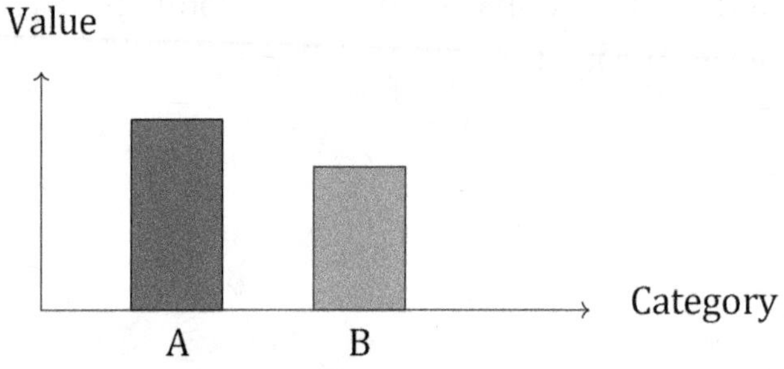

Figure 5.10: Bar graph with an exaggerated y-axis. Significant differences appear when the graph is cut off below 1.

Always be cautious of graph scales and make sure to check the full range of the axes to avoid misinterpretations.

Conclusion

In summary, understanding range and variability, how data is distributed, and how to recognize misleading data presentations will significantly boost your data interpretation skills. Prepare to see questions related to these concepts in the GED Math Test, as they form the core of analyzing and interpreting numerical data.

Exercises

Exercise 5.12.

1. Given the data set: 12, 15, 22, 13, 19, 17, 20, 23, calculate the range of the data.

2. Calculate the interquartile range (IQR) for the following data set: 3, 7, 8, 5, 12, 14, 21, 13, 18, 19.

3. A restaurant recorded the number of customers per day for a week: 120, 135, 140, 130, 125, 155, 145 Identify and plot the data distribution. What is the mode of the data set?

Exercise 5.13.

Consider the following chart showing the distribution of test scores in a math exam. Study the chart and:

1. Describe the shape of the data distribution.

2. Identify if the data is skewed left, skewed right, or symmetric.

3. Suggest what might cause this type of distribution (e.g., test difficulty, student preparedness).

Exercise 5.14.

A survey result is plotted below using a bar graph (assume bars represent number of participants across different age groups). Explore the data and:

1. Determine if any bar appears misleading.

2. Suggest ways the graph can be improved for clearer data interpretation.

3. Explain how visual misrepresentation in the graph could affect the way results are reported.

Answers with Explanations

Solution 5.12.

1. The range of a data set is the difference between the maximum and minimum values. In the given data set 12, 15, 22, 13, 19, 17, 20, 23:

$$\text{Range} = 23 - 12 = 11.$$

Therefore, the range is 11.

2. To calculate the interquartile range (IQR), we first organize the data:

$$3, 5, 7, 8, 12, 13, 14, 18, 19, 21.$$

The median points are 12 and 13, and the lower quartile (Q1) is the median of the lower half, 7, while the upper quartile (Q3) is the median of the upper half, 18:

$$IQR = Q3 - Q1 = 18 - 7 = 11.$$

Therefore, the IQR is 11.

3. The mode is the most frequent value. Since each customer count in the given data 120, 135, 140, 130, 125, 155, 145 appears once, this data has no mode. A bar graph of this data, labeled as "Day" on the x-axis and "Number of Customers" on the y-axis, would show relatively even distribution with no dominant value point (no mode).

Solution 5.13.

The given chart shows a distribution of test scores: 30

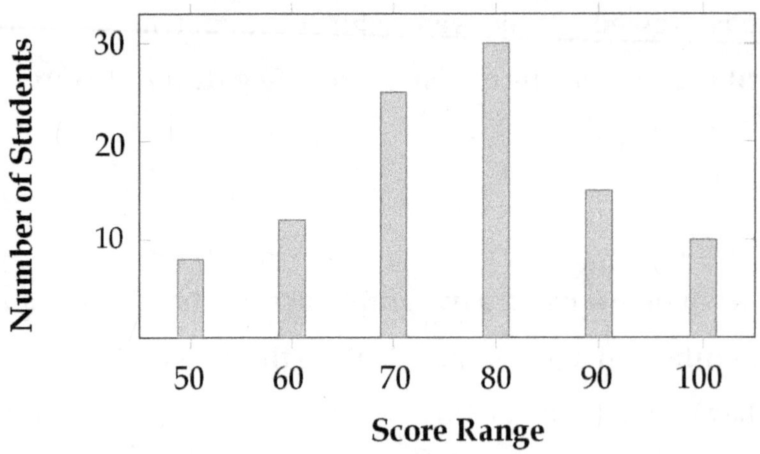

1. The distribution shows most scores are concentrated between 70 and 90, and fewer scores are at the extremes. This is an example of a normal distribution.

2. The distribution appears slightly skewed to the left as there are more students scoring at the lower end of the tail, such as 50 and 60.

3. This skew could occur if the test was challenging for many students, creating a grouping near the middle of the score range with fewer high scorers.

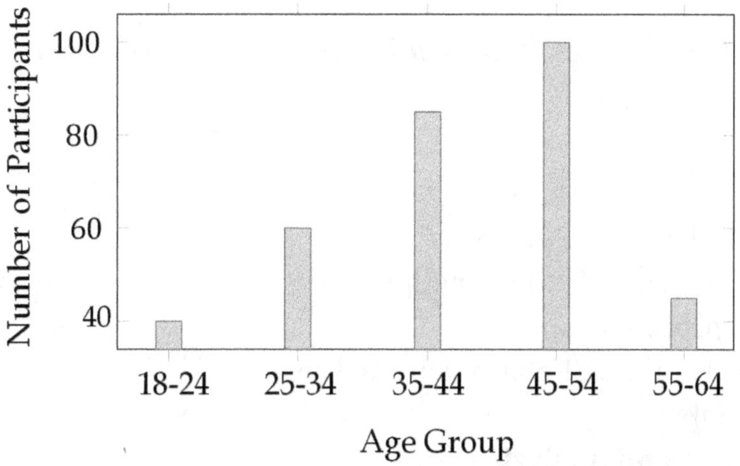

Solution 5.14.

1. The bar for the age group 45 – 54 seems misleading because it is significantly larger than the others. The y-axis might have been manipulated to exaggerate the difference, i.e., its unit scale may vary.

2. The graph could be improved by ensuring a consistent y-axis scale, splitting the data into smaller intervals for more detail, and avoiding truncated or exaggerated heights.

3. Misleading graphs can exaggerate or minimize certain trends, potentially leading to false conclusions. For example, in this case, the reader might believe the majority of participants are from the 45 – 54 group, when in fact the difference is not that extreme.

CHAPTER # 6

Graphical and Functional Representation

Math topics that you'll learn in this chapter:

- Graphs of Functions
- Plotting Linear Functions
- Graphing Quadratic Functions
- Interpreting Graphs
- Understanding Relationship Between Variables
- Reading and Analyzing Graphs
- Function Transformations
- Slope and Intercept
- Concept of Slope

6.1 Graphs of Functions

Graphing functions is one of the most fundamental concepts in algebra and can help visualize the relationship between variables. In this section, we will explore how to plot linear and quadratic functions, and how to identify important features such as intercepts and the vertex.

6.1.1 Plotting Linear Functions

A linear function is a function of the form:

$$f(x) = mx + b$$

where m is the slope, and b is the y-intercept. Let's break down how we can plot this function step by step.

Example 1:
Plotting

$$f(x) = 2x + 3$$

1. Identify the slope (m) and intercept (b): The slope $m = 2$ and the y-intercept $b = 3$. This tells us that the line will pass through the point (0, 3) and will increase by 2 units for every 1 unit increase in x.

2. Plot the intercept: Start by plotting the point where the line crosses the y-axis, which is at (0, 3).

3. Use the slope: The slope $m = 2$ means the line rises 2 units for every 1 unit it moves to the right. From (0, 3), move one unit right (to $x = 1$) and move up 2 units (to $y = 5$). Plot this new point at (1, 5).

4. Draw the line: Use a ruler to connect the points (0, 3) and (1, 5), then extend the line in both directions.

This gives us the plot of the linear function $f(x) = 2x + 3$.

Exploring Slope

The slope m controls the steepness of the line. Let's look at how different values of m affects the graph:

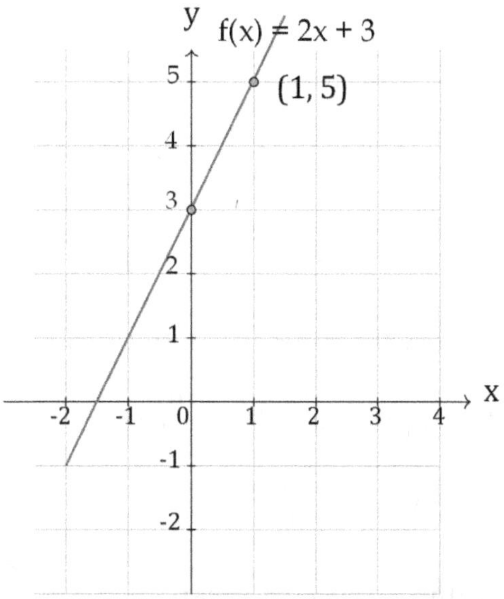

Figure 6.1: Graph of Linear Function f(x) = 2x + 3

1. Positive Slope ($m > 0$): If the slope is positive, such as in the function $f(x) = 2x + 3$ the line rises as it moves from left to right.

2. Negative Slope ($m < 0$): If the slope is negative, the line will fall as it moves from left to right (e.g., $f(x) = -2x + 3$).

3. Zero Slope ($m = 0$): If the slope is zero, the function represents a horizontal line (e.g., $f(x) = 3$).

4. Undefined Slope (Vertical Line): A vertical line has an undefined slope, and it is not represented by the equation $f(x) = mx + b$, but instead with an equation like $x = a$, where a is a constant.

6.1.2 Graphing Quadratic Functions

A quadratic function can be written in the general form:

$$f(x) = ax^2 + bx + c$$

The graph of a quadratic function is a parabola. Let's look at how to graph quadratic functions and identify key features like the vertex and intercepts.

Example 2:

Plotting $f(x) = x^2 - 4x + 3$

1. Identify the vertex: The vertex of a parabola given by $f(x) = ax^2 + bx + c$ is found using

the formula for the x-coordinate:

$$x = -\frac{b}{2a}$$

In our example, a = 1, b = −4, and c = 3. Using the formula:

$$x = -\frac{-4}{2(1)} = 2$$

To find the y-coordinate of the vertex, substitute x = 2 into the quadratic function:

$$y = f(2) = (2)^2 - 4(2) + 3 = 4 - 8 + 3 = -1$$

Thus, the vertex is at (2, −1).

2. Find the y-intercept: The y-intercept occurs when x = 0. Substituting x = 0 into f(x):

$$f(0) = (0)^2 - 4(0) + 3 = 3$$

So, the y-intercept is at the point (0, 3).

3. Find the x-intercepts: The x-intercepts occur where f(x) = 0, so solve:

$$x^2 - 4x + 3 = 0$$

Factoring the quadratic expression:

$$(x - 3)(x - 1) = 0$$

This gives x = 3 and x = 1. Therefore, the x-intercepts are at (1, 0) and (3, 0).

4. Plot and Draw: Plot the vertex (2, −1), the y-intercept (0, 3), and the x- intercepts (1, 0) and (3, 0). Then sketch the parabola passing through these points.

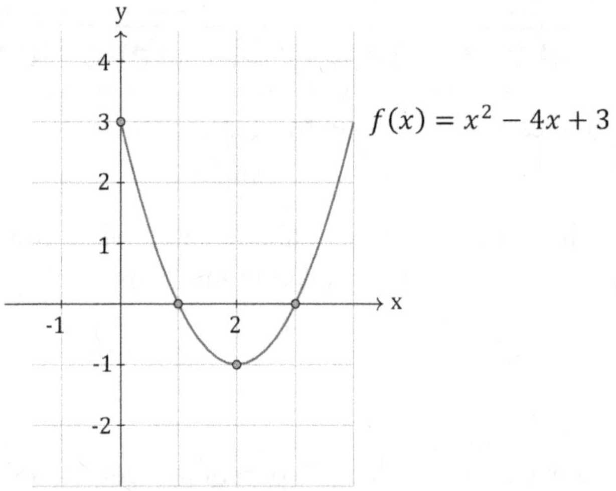

Figure 6.2: Graph of Quadratic Function $f(x) = x^2 - 4x + 3$

Identifying Key Features

When graphing any quadratic function, identifying certain key features can aid in the graphing process:

1. Vertex: The vertex represents the highest or lowest point of the parabola. It is either a maximum or minimum depending on whether the parabola opens upwards ($a > 0$) or downwards ($a < 0$).
2. Y-Intercept: This is the point where the graph crosses the y-axis and occurs when $x = 0$.
3. X-Intercepts: These are the points where the graph crosses the x-axis. They can be found by solving $f(x) = 0$.

In the next subsections, we will explore more about the transformations of these functions and how the graphs change when we modify certain parameters. This under- standing will enhance your graph interpretation skills for both linear and quadratic equations.

Exercises

Exercise 6.1.

Plot the linear function $f(x) = 2x + 3$ for x $[-5, 5]$. Label the intercepts on the graph.

Exercise 6.2.

Graph the quadratic function $f(x) = -x^2 + 4x - 1$ and determine the vertex and intercepts.

Exercise 6.3.

Identify the key features (including intercepts and vertex) of the quadratic function $f(x) = -2x^2 + 4x + 6$. Plot the function and label the key features.

Answers with Explanations

Solution 6.1.

The linear function is $f(x) = 2x + 3$. To plot the function, we need to find both the y-intercept (point where the function crosses the y-axis) and the x-intercept (point where the function crosses the x-axis).

The y-intercept is found by setting $x = 0$:

$$f(0) = 2(0) + 3 = 3$$

So, the y-intercept is at $(0, 3)$.
The x-intercept can be found by setting $f(x) = 0$:

$$0 = 2x + 3 \Rightarrow x = -\frac{3}{2}$$

So, the x-intercept is at $\left(-\frac{3}{2}, 0\right)$

Now, we plot the function. Here's the graph of $f(x) = 2x + 3$:

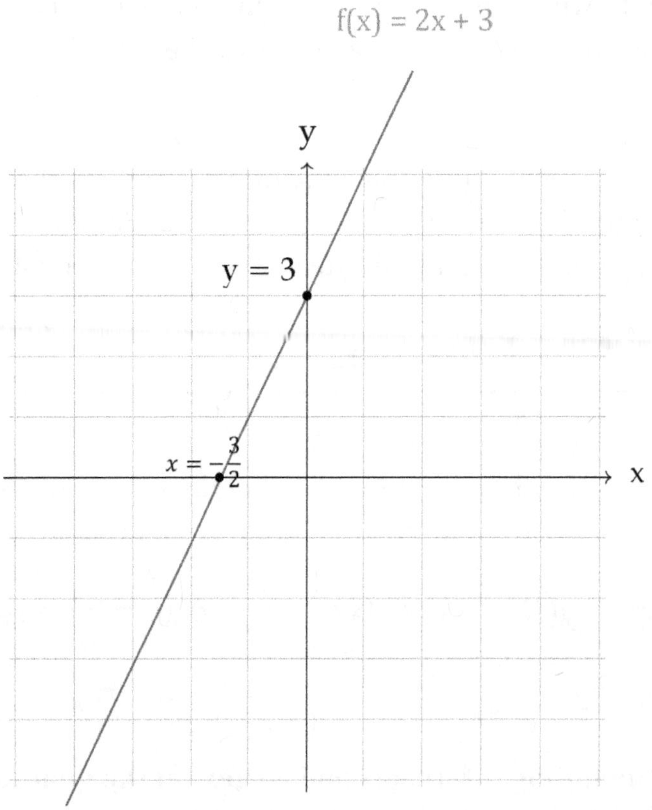

Solution 6.2.

The quadratic function is $f(x) = -x^2 + 4x - 1$.

To find the vertex, we use the formula for the x-coordinate of the vertex:

$$x = \frac{-b}{2a}$$

For $f(x) = -x^2 + 4x - 1$, $a = -1$ and $b = 4$:

$$x = \frac{-4}{2(-1)} = 2$$

To find the y-coordinate of the vertex, we substitute $x = 2$ into the equation:

$$f(2) = -(2)^2 + 4(2) - 1 = -4 + 8 - 1 = 3$$

So, the vertex is at $(2, 3)$.

The x-intercepts are found by solving $-x^2 + 4x - 1 = 0$ using the quadratic formula:

$$x = \frac{-b \pm \sqrt{b^2 - 4ac}}{2a}$$

Substituting $a = -1$, $b = 4$, and $c = -1$:

$$x = \frac{-4 \pm \sqrt{4^2 - 4(-1)(-1)}}{2(-1)} = \frac{-4 \pm \sqrt{16 - 4}}{-2} = \frac{-4 \pm \sqrt{12}}{-2}$$

Simplifying:

$$x = \frac{-4 \pm 2\sqrt{3}}{2} = 2 \pm \sqrt{3}$$

The approximate values are $x \approx 2 - 1.73 \approx 0.27$ and $x \approx 2 + 1.73 \approx 3.73$.

The y-intercept is found by setting $x = 0$:

$$f(0) = -0^2 + 4(0) - 1 = -1$$

So, the y-intercept is at $(0, -1)$.

Now, we plot the graph:

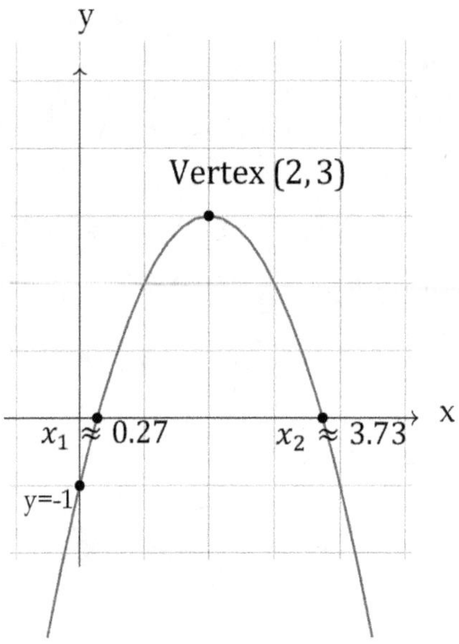

Solution 6.3.
The quadratic function is $f(x) = -2x^2 + 4x + 6$.
To find the vertex

$$x = \frac{-b}{2a} = \frac{-4}{2(-2)} = 1$$

Substituting x = 1 into f(x):

$$f(1) = -2(1)2 + 4(1) + 6 = -2 + 4 + 6 = 8$$

So, the vertex is at (1, 8)
Solving for the x-intercepts using the quadratic formula:

$$x = \frac{-4 \pm \sqrt{4^2 - 4(-2)(6)}}{2(-2)} = \frac{-4 \pm \sqrt{16 + 48}}{-4} = \frac{-4 \pm \sqrt{64}}{-4}$$

$$x = \frac{-4 \pm 8}{-4}$$

The two solutions are x = 3 and x = −1.

The y-intercept is found by setting x = 0

$$f(0) = -2(0)2 + 4(0) + 6 = 6$$

So, the y-intercept is (0, 6).

The graph is shown below:

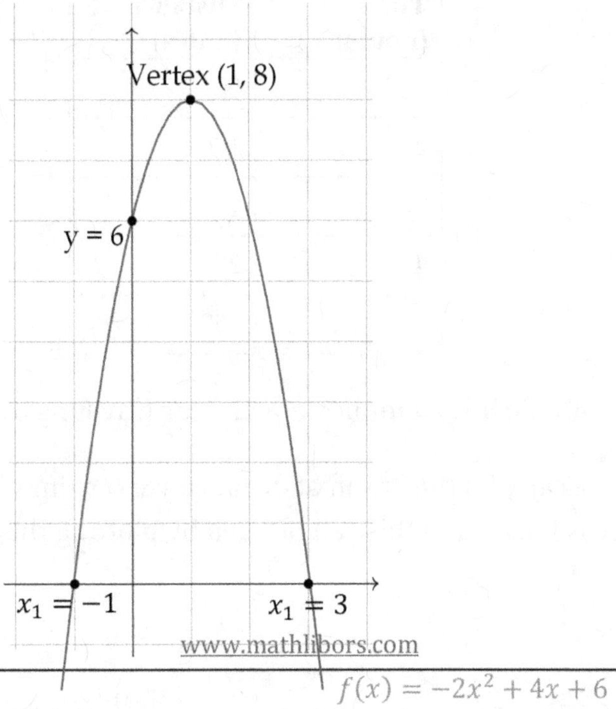

6.2 Interpreting Graphs

Graphs are powerful tools that allow us to visualize relationships between variables. In the context of GED math, interpreting graphs is an essential skill, particularly when analyzing functions or understanding data from real-life situations. This section will cover three key aspects: understanding the relationship between variables, reading and analyzing basic graphs, and interpreting function transformations.

6.2.1 Understanding Relationship Between Variables

Graphs often illustrate how two variables are related to one another. The most common setup is the use of a Cartesian coordinate plane (or graph), where the horizontal axis (also known as the x-axis) typically represents the input or independent variable, and the vertical axis (the y-axis) commonly represents the output or dependent variable.

For example, consider a scenario where you are tracking the distance a car travels as time passes. In this case:

- The independent variable (x) represents **time**.
- The dependent variable (y) represents **distance**.

Example: Distance-Time Relationship
Let's imagine that we have the following data about a car traveling at a constant speed:

Time (hours)	**Distance (miles)**
1	50
2	100
3	150
4	200
5	250

Table 6.1: Example data of a car traveling over time.

Using this table, we can plot the graph of distance versus time. The relationship between time and distance is linear, and this can be seen by plotting the points (1, 50), (2, 100), (3, 150) and so on.

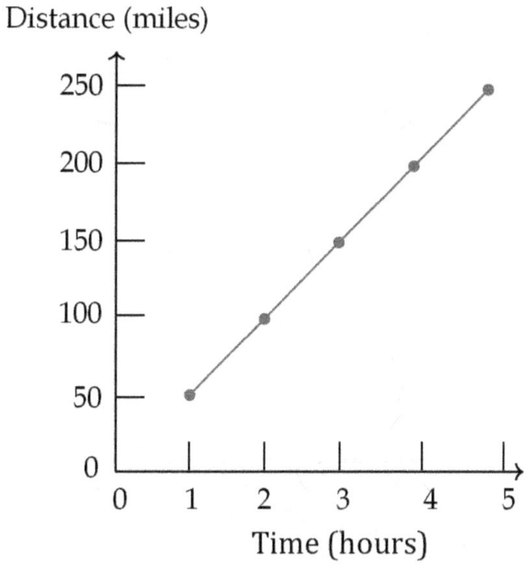

Figure 6.3: Graph of Time vs. Distance for a car traveling at a constant speed.

In the example above, we observe that the graph represents a **linear relationship** between time and distance. The slope of the line tells us that the car is traveling at a constant speed (50 miles per hour), which is simply the rate of change of distance with respect to time.

6.2.2 Reading and Analyzing Graphs

When interpreting a graph, it is essential to understand the type of graph being used and the kind of information it conveys. Basic graphs that you will encounter include:

- Line graphs: Often used to show relationships between two variables over time.
- Bar graphs: Useful for comparing quantities across different categories.
- Pie charts: Display how a whole is divided into various parts.

Example: Reading a Temperature-Time Line Graph

Consider the following example: a line graph depicting the temperature at different hours in a day.

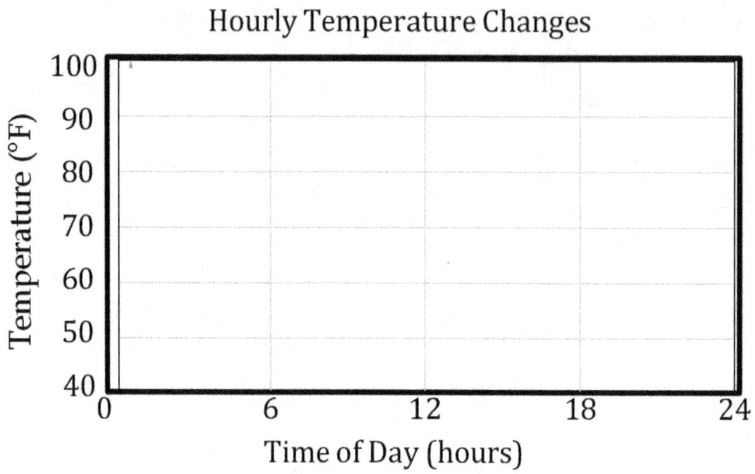

Figure 6.4: Temperature change over a 24-hour period.

In this graph:

- **Time of Day** is plotted along the x-axis.
- **Temperature** is on the y-axis. From the graph, observe that:
- At midnight (0 hours), the temperature was 60°F.
- The temperature dropped to 50°F at 6 a.m.
- By 12 p.m., it reached 80°F.
- At 6 p.m., the temperature peaked at 90°F.
- After that, the temperature slightly decreased to 70°F by midnight again.

Such graphs allow us to analyze how variables change over time and identify trends, such as minimum and maximum temperatures over a day.

6.2.3 Function Transformations

Graphs of functions can be modified by performing transformations. These transformations include translating (shifting), reflecting, stretching, and compressing the graph. Let's examine each transformation type with examples.

1. Vertical and Horizontal Shifts A **vertical shift** occurs when a constant is added or subtracted to the entire function:

$$y = f(x) + c$$

Adding c shifts the graph up by c units, while subtracting c shifts it down by c units.

A "horizontal shift" occurs when a constant is added or subtracted inside the function's argument:

$$y = f(x + h)$$

Adding h shifts the graph to the left by h units, and subtracting shifts it to the right by h units.

Example: Shifting the Graph of $y = x^2$ Consider the graph of the function $y = x^2$:

$$y = (x)^2 \text{ (Original graph)}$$

$$y = (x - 2)^2 \text{ (Shifted right by 2 units)}$$

$$y = x^2 + 3 \text{ (Shifted up by 3 units)}$$

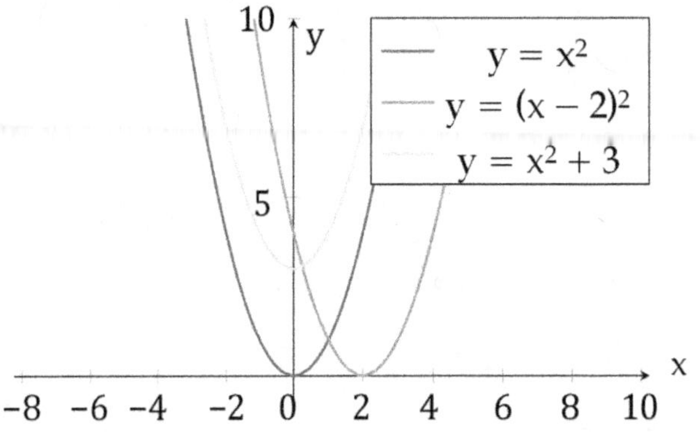

Figure 6.5: Shift transformations on the parabola $y = x^2$.

2. Reflections Reflecting a graph across the x or y axis involves multiplying by −1:

$$y = -f(x) \quad \text{(reflection across the x-axis)}$$
$$y = f(-x) \quad \text{(reflection across the y-axis)}$$

3. Stretching and Compressing Graphs Scaling transformations change the shape of the graph by stretching or compressing it vertically or horizontally:

$$y = a \cdot f(x) \text{ (Stretch/compress vertically by factor a)}$$
$$y = f(kx) \text{ (Stretch/compress horizontally by factor k)}$$

Each of these transformations modifies the appearance of the function's graph but retains the core relationship between the variables.

In GED math, tasks involving interpreting function graphs are common, including identifying key features like intercepts, slope, and transformations. By practicing with real-world applications and function graphs, you will strengthen your ability to not only read graphs but also analyze transformations and understand how different variables interact visually.

Exercises

Exercise 6.4.

Consider the graph below showing the relationship between hours studied (x) and score achieved on a test (y). Answer the following questions:

1. Identify the type of relationship between the two variables. Is it linear, quadratic, or some other form?

2. Estimate the score corresponding to 3 hours of study from the graph.

3. Explain whether an increase in hours studied always results in a higher score, based on the trend in the graph.

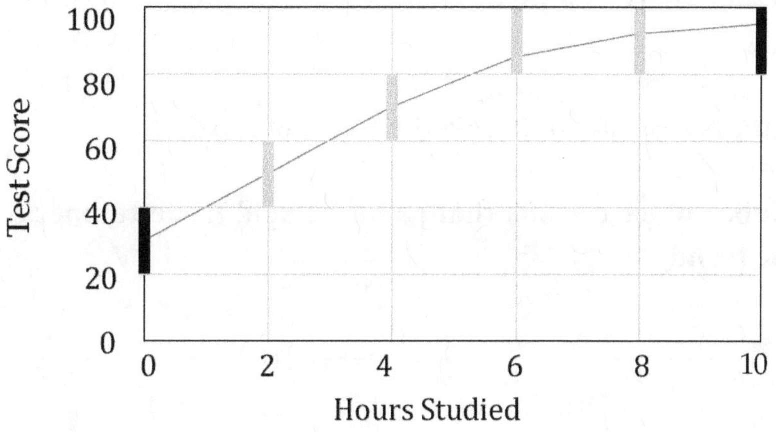

Exercise 6.5.

Examine the bar chart below that compares the number of books read in different months.

1. Which month had the highest number of books read?

2. What is the total number of books read across all 4 months?

3. Compare the number of books read in February to the number read in April. By what percent did the number of books increase or decrease?

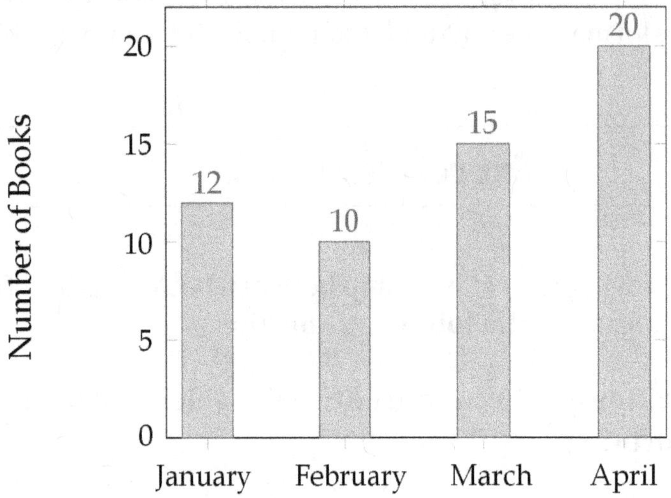

Exercise 6.6.

The scatter plot below shows the relationship between temperature (in degrees Celsius) and the number of ice creams sold.

1. Determine whether there is a positive, negative, or no correlation between the variables.

2. Identify any outliers that do not follow the trend in the data.

3. Predict the number of ice creams that could be sold if the temperature reaches 30 °C, based on the trend.

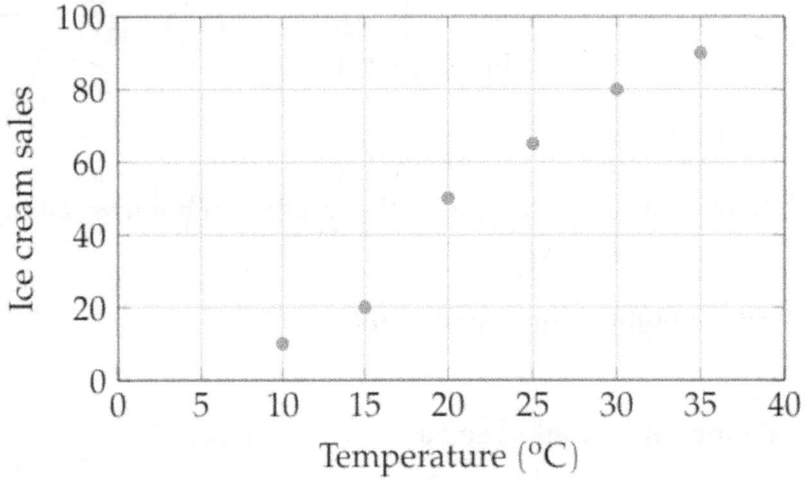

Answers with Explanations

Solution 6.4.

1. The relationship between hours studied (x) and score (y) appears to be approximately linear, as the points on the graph follow a straight-line pattern.
2. From the graph, the score corresponding to 3 hours of study is approximately 60.
3. Yes, an increase in hours studied generally results in a higher score, but the increment diminishes at the higher end (after 8 hours), suggesting diminishing returns.

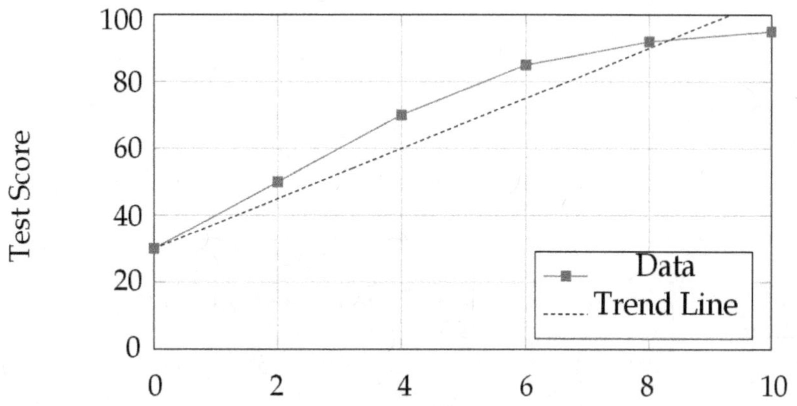

Solution 6.5.

1. April had the highest number of books read, with 20 books.
2. The total number of books read across all the months is $12 + 10 + 15 + 20 = 57$.
3. The number of books read in February is 10, and in April, it is 20. The percent increase is:

$$\frac{20 - 10}{10} \times 100 = 100\%$$

Thus, the number of books read doubled from February to April.

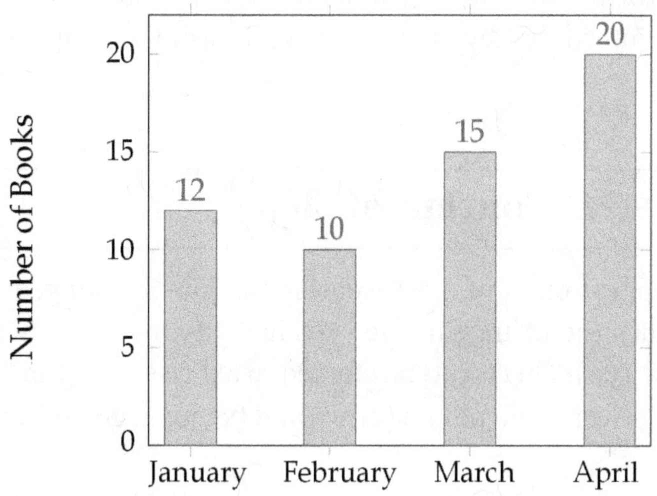

Solution 6.6.

1. The relationship between temperature and ice cream sales demonstrates a positive correlation, meaning that as the temperature increases, ice cream sales also increase.

2. There do not appear to be any significant outliers that deviate from the trend.

3. If the temperature reaches 30°C, the number of ice creams sold could be predicted to be approximately 80, based on the trend.

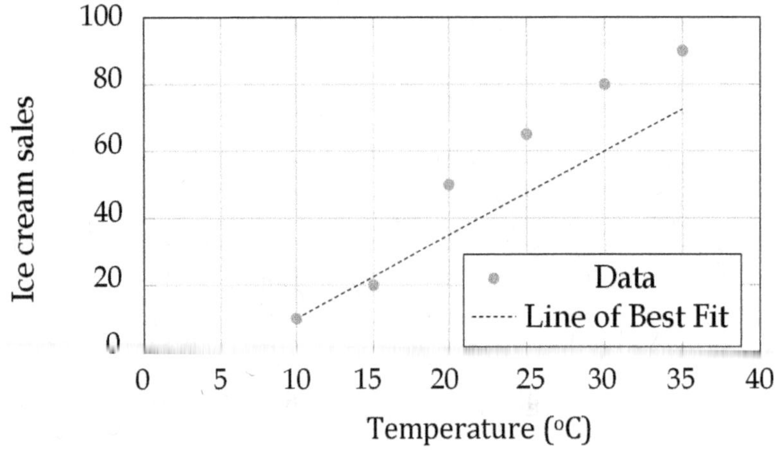

6.3 Slope and Intercept

In this section, we will delve into two essential components of linear equations: the slope and the intercept. These ideas form the foundation for understanding the properties and behavior of linear functions. We will break down the concepts into three subsections—understanding the slope, finding the y-intercept, and using the slope- intercept form to write and interpret equations of lines. By the end of this section, you should be able to write the equation of a line and interpret key details about it from a graph or algebraic expression.

6.3.1 Concept of Slope

The **slope** measures the steepness of a line, representing the rate of change between two points. Specifically, it describes the change in the "y-value" (Δy, or rise) divided by the change in the "x-value" (Δx, or run). This is often referred to as "rise over run."
Mathematically, the slope m between two points (x_1, y_1) and (x_2, y_2) is given by the formula:

$$m = \frac{y_2 - y_1}{x_2 - x_1}$$

Example: Suppose you are given two points, A $(2,3)$ and B $(5,7)$. To find the slope of the line connecting these points, apply the formula:

$$m = \frac{7-3}{5-2} = \frac{4}{3}$$

So, the slope of the line is $\frac{4}{3}$, meaning for every 3 units you move horizontally, the line rises 4 units vertically.

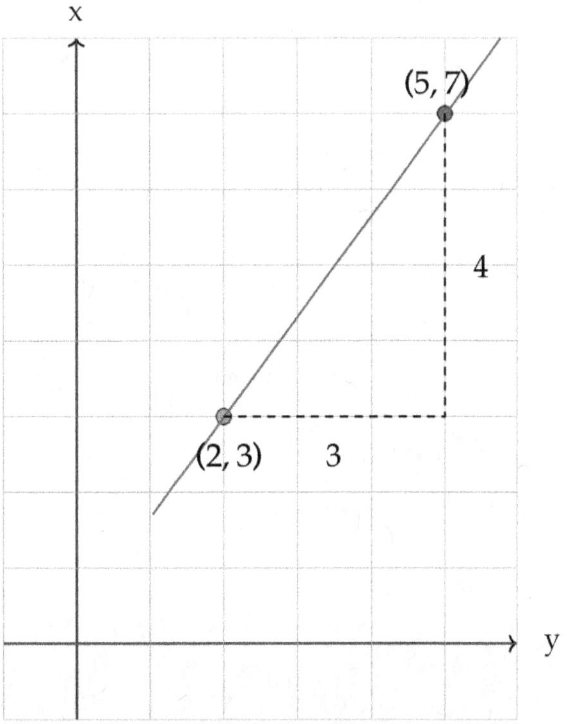

Figure 1: Plot of two points A $(2, 3)$ *and* B $(5, 7)$ *with the triangle showing slope* $\frac{4}{3}$.

The slope can be positive, negative, zero, or undefined, depending on how the line rises or falls:

- Positive Slope: The line rises from left to right.
- Negative Slope: The line falls from left to right.
- Zero Slope: The line is horizontal.
- Undefined Slope: The line is vertical.

Example:

Positive slope: m = 2 (The line moves up as you go from left to right)
Negative slope: m = −3 (The line moves down as you go from left to right)
Zero slope: m = 0 (The line is horizontal)
Undefined slope: (The line is vertical; no slope exists)

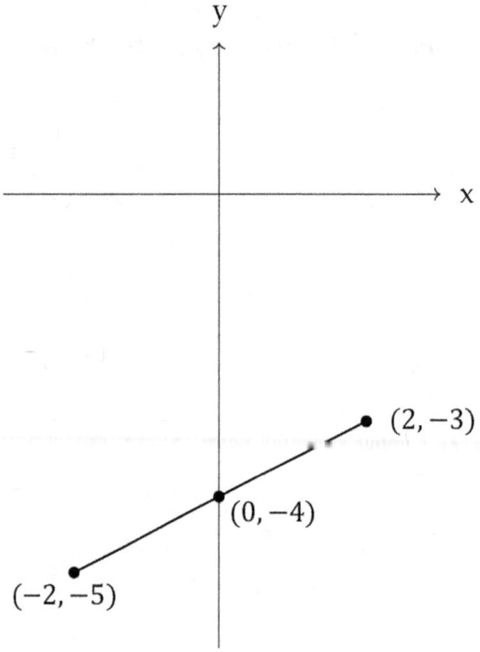

CHAPTER # 7

Math Test-Taking Strategies

Math topics that you'll learn in this

- Time Management
- Pacing Yourself
- When to Skip Questions
- Using Remaining Time Wisely
- Answering Multiple Choice Questions
- Process of Elimination
- Guessing Strategies
- Reviewing Your Work
- Double-Checking Units
- Confirming Calculations

7.1 Time Management

Time management is a crucial skill to master when taking the GED Math test. Since the test is timed, effective time allocation can reduce stress, improve performance, and ensure you answer as many questions as possible. In this section, we will explore three key strategies for managing your time: pacing yourself, knowing when to skip questions, and using remaining time wisely.

7.1.1 Pacing Yourself

One of the most important aspects of time management on the GED Math test is ensuring that you pace yourself correctly. Rushing may lead to careless mistakes, while spending too much time on one problem can cost you valuable opportunities to answer easier questions later in the test.

The GED Math test consists of approximately 45 questions, and you are given 115 minutes (1 hour and 55 minutes) to complete the test. This means you have just over 2.5 minutes per question on average. However, not all questions are equally difficult, so a crucial part of pacing yourself is recognizing question difficulty and adjusting your time accordingly. Let's break this down:

$$\frac{115\ minutes}{45\ questions} \approx 2.56\ minutes\ per\ questions$$

For easier questions, you should be able to answer them quickly, ideally in less than one minute. On the other hand, more challenging questions may take longer. Managing where you spend your time is important for ensuring you get through all of the questions.

Example 1:

Consider the following problem, which is relatively simple:

$$What\ is\ 3 + 4 \times 2?$$

Using the *Order of Operations* (or PEMDAS), we know to multiply before adding:

$$4 \times 2 = 8,$$
$$3 + 8 = 11.$$

This question should take under 30 seconds to solve. By quickly recognizing it as an easy problem, you can complete it quickly and move on to tougher questions, conserving time for when you really need it.

7.1.2 When to Skip Questions

Another important skill in time management is knowing when to skip a question. There will be some questions that are particularly tricky or time-consuming. Spending too long on a single problem can drain your time and prevent you from answering multiple questions you might otherwise get correct.

Rule of Thumb: If you find yourself stuck on a problem for more than 2-3 minutes, it may be time to skip it and return to it later.

Skipping questions doesn't mean abandoning them forever. The key is to mark the question and come back to it with a fresh perspective after you've tackled the easier questions.

Example 2:
Suppose you encounter a more complex algebraic equation:

$$2(x-3)+3x=4(x+1)-5.$$

If you're struggling to figure it out within the standard time (2-3 minutes), it's better to flag this question, move on, and come back to it later. Panicking and getting stuck on this question may prevent you from answering multiple easier questions down the line.

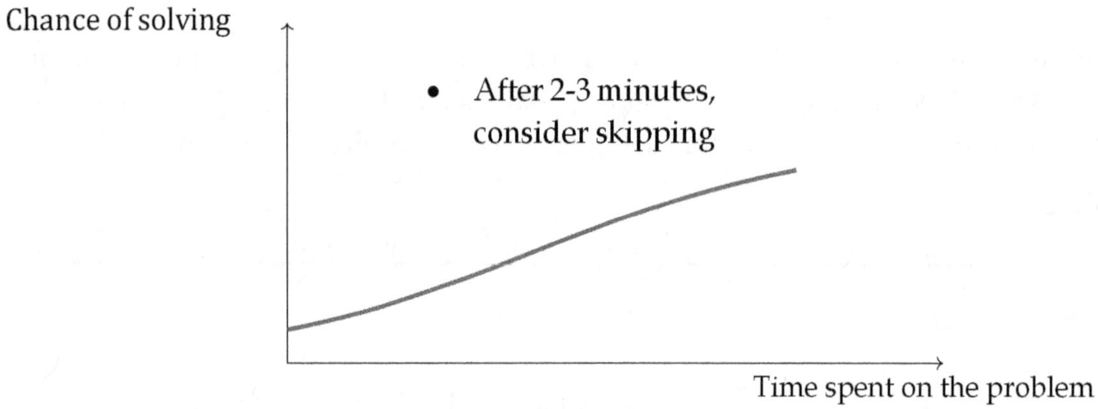

Figure 7.1: Time spent vs. chance of solving a difficult problem.

7.1.3 Using Remaining Time Wisely

Once you've completed a first pass through the test, it's important to use any remaining time wisely. At this point, you should return to the questions you initially skipped or felt unsure about. This approach ensures you've maximized the number of questions you've attempted without getting stuck prematurely.

Techniques for Revising in the Remaining Time:

1. Review flagged questions: Go back to those challenging problems you skipped. You may have a clearer mind or fresh insight that now helps you solve them more easily.

2. Double-check your answers: Use the extra time to check over your work, especially in problems involving multi-step calculations.

3. Simple spot mistakes: Ensure you've followed the proper order of operations and confirm you didn't misread any questions.

Example 3:

You have 5-10 minutes left in the test, and there are three questions you previously skipped. Now is the time to revisit them:

- One of the skipped questions is a word problem involving ratios, asking you to determine the correct ratio of students in different study groups. With extra time, you reread the problem slowly and realize that you initially misinterpreted one part. You're able to adjust your approach and solve it in 2 minutes.

Finally, after all skipped questions have been addressed, use any further time to make sure you've bubbled in your answers correctly on the answer sheet, and review questions where you guessed earlier.

Time Remaining	Action Steps
10+ minutes	Flagged questions should be revisited first. Spend time systematically solving any skipped questions, especially those you now feel more confident about.
5-10 minutes	Start reviewing questions you guessed. Confirm calculations, look for simple errors like addition mistakes, and verify you've fully answered the problem. Make sure your answers are properly marked on your answer sheet
Less than 5 minutes	Prioritize correcting any noticeable mistakes. Use this phase to focus on high-value sections where errors are likely, like multistep algebraic problems or geometrical constructions. Don't try to redo entire problems; focus on small adjustments

Table 7.1: Checklist for Effective Time Use Once You're Finishing the Test

By mastering these time management techniques—pacing yourself, skipping questions strategically, and efficiently using remaining time—you can maximize your chances of achieving a high score on the GED Math test.

Exercises

Exercise 7.1.

During a practice test, you notice there are 50 multiple-choice questions, and you only have 75 minutes to complete the test. How much time should you spend on each question on average to pace yourself and finish the test on time?

Exercise 7.2.

While working on a timed math test, you encounter a difficult problem that you are unsure how to solve. Describe how you would approach this situation to ensure you don't spend too much time on this question.

Exercise 7.3.

When you have completed a timed test with 10 minutes remaining, what strategies can you use to ensure that you are fully utilizing the remaining time wisely before submitting?

Answers with Explanations

Solution 7.1.

To pace yourself, you need to divide the total time evenly across all the questions. You have 50 questions and 75 minutes.

$$Time\ per\ question = \frac{75\ minutes}{50\ questions} = 1.5\ minutes\ per\ question.$$

Therefore, you should spend approximately 1 minute and 30 seconds on each question to finish the test on time.

Tip: Use a watch or the test timer to check your progress after every few questions to ensure you're staying within the time limit.

Solution 7.2.

If you encounter a difficult problem, you should consider skipping the question and returning to it later. Here's one approach:

1. Mark the question for review.

2. Quickly move to the next question.

3. After answering the remaining questions, return to the marked question if time allows.

This strategy ensures that you don't waste too much time on a difficult question while leaving easier questions unanswered.

Helpful tip: Sometimes, working on other questions may give you new insights into the difficult one when you return to it.

Solution 7.3.

To effectively use the remaining 10 minutes of the test, consider the following strategies:

1. Review flagged or skipped questions first. These are questions that you were unsure about or didn't have time to answer thoroughly.

2. Double-check your answers for any calculation errors or misinterpretations of the question.

3. Ensure that all questions are answered, even if it requires making an educated guess for difficult ones.

4. Confirm that your answers in the answer sheet or digital entry match correctly with the questions, particularly if you had jumped between questions.

Diagram: Below is an illustration showing a timeline for the last 10 minutes of the test review process:

Timeline for Last 10 Minutes

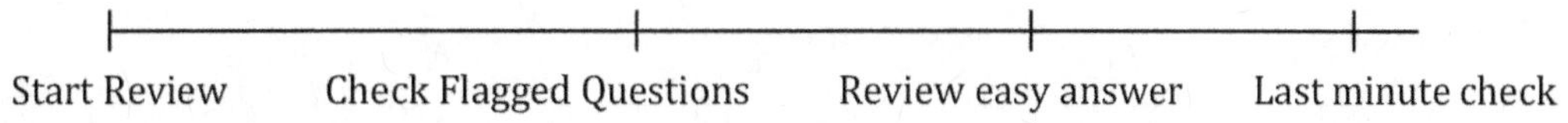

This approach minimizes the likelihood of leaving easy points on the table while ensuring that difficult questions are reevaluated.

7.2 Answering Multiple Choice Questions

In this section, we will delve into efficient strategies for mastering multiple-choice questions on the GED Math test. Answering multiple-choice questions correctly requires more than just knowledge of the subject matter; it also involves strategies like the process of elimination, educated guessing, and reviewing your work systematically.

These approaches can help improve your chances of success and enhance your test-taking confidence.

7.2.1 Process of Elimination

The process of elimination is a powerful technique for test-takers. Even if you don't immediately know the correct answer, narrowing down the options will significantly increase your chances of guessing correctly. Follow these steps:

- **Eliminate clearly wrong answers**: Start by quickly scanning the answer choices and crossing out those that are obviously wrong.

- **Analyze the remaining choices**: Carefully read the question again, paying close attention to details, and then analyze the remaining options. Try to identify patterns or features that make one choice stand out as more likely correct.

- **Look for "trap" answers**: Test designers often include distractors that seem correct but are not. Common traps might include answers that involve common math mistakes, such as errors in arithmetic, misapplication of formulas, or misconceptions about key concepts.

Example:
Suppose a multiple-choice question asks you to solve the following equation:

$$2x + 5 = 17$$

The available answer choices are:

A. 6

B. 7

C. 5

D. 4

Follow the process of elimination:

1. You can begin by solving the equation: subtract 5 from both sides,

$$2x = 12$$

and then divide by 2 to get:

$$x = 6$$

2. You immediately know that the answer is **A.** 6. If you somehow couldn't solve the equation but had to guess, notice that the other options might seem plausible but involve common calculation traps (e.g., forgetting certain steps or miscalculating).

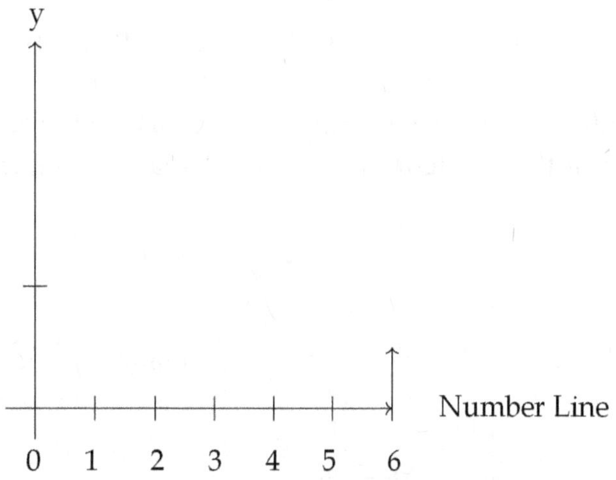

x=6 is the correct answer

Figure 7.2: Solving Equation $2x + 5 = 17$ using number line

7.2.2 Guessing Strategies

When you're unsure of an answer and elimination doesn't work completely, guessing can be your last resort. However, using informed guessing strategies can skew the odds in your favor.

- **Never leave a question blank**: There is no penalty for guessing on the GED Math test, so make sure to always put an answer. A random guess still has a 25% chance of being correct.

- **Look for patterns**: If certain options break patterns (e.g., three answers have similar numbers, and one is drastically different), the outlier may often be a red herring or is very likely to be right.

- **Check for units**: Make sure to check that the answer choices have the correct units (e.g., feet vs. inches) that might be asked in the question. This will help you eliminate answers that don't match the required units.

- **Use logical reasoning**: Get rid of extreme values. If two answers are large and two are much smaller, often the large values might be there to trap those who made a specific mistake in the process of calculation.

Example:
If you're solving for the area of a triangle, and the question gives options like:

A. 12 units^2, **B.** 5 units^2, **C.** 50 units^2, **D.** 1000 units^2

Looking at the drastic difference between **D** and the rest of the choices, it's clear that **D** is likely either a distractor or the result of a gross miscalculation. You can eliminate **D** in this case.

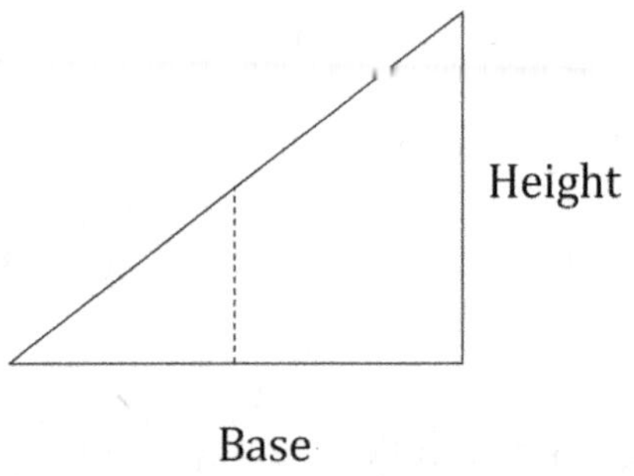

Figure 7.3: Identifying base and height of the triangle

7.2.3 Reviewing Your Work

Reviewing your answers before submitting your exam is an essential step in avoiding careless mistakes. Follow these tips when reviewing your work:

- **Double-check calculations**: Especially when using a calculator, re-enter the numbers to ensure accuracy. Check steps in complex equations to ensure no steps were skipped.

- **Ensure you answered what was asked**: Many questions trap students by providing intermediate numbers that look like the answer. Reread the question to make sure you're answering it entirely.

- **Check for common errors**: Look for units, signs (positive vs. negative), and decimals that might affect your final choice.

- **Change answers wisely**: Only change an answer if you find a definitive reason to do so during review. Often, your first instinct is correct unless you identify a clear mistake.

Example:

Consider this word problem:

"A rectangle has a width of 5 units and an area of 40 square units. What is the length of the rectangle?"

After solving it, you might get an answer of L = 8 units.

The answer choices are:

A.5 units

B.8 units

C.7 units

D.40 units

If you weren't careful, you might mistakenly choose **D** because it matches the number in the problem (the area), but the question asks for *length*. Always re-read the question and ensure that you're providing the right numerical answer.

Step	Correct Process	Common Mistake
Step 1: Area Formula Step 2: Plug in known values Step 3: Solve for L	A = L x W 40 = L x 5 L = 40/5 = 8	Forgetting to apply the area formula Misinterpreting the given numbers Accidentally using area value for the wrong variable

Table 7.2: Process for solving area-related word problems

By mastering these approaches - eliminating obvious wrong answers, using proven guessing strategies, and reviewing your work thoroughly - you can significantly improve your accuracy and efficiency when answering multiple-choice questions on the GED Math test.

Exercises

Exercise 7.4.

Using the process of elimination, which of the following is the best answer to the question: "What is the square root of 81?"

1. **7**
2. **9**
3. **12**
4. **10**

Briefly explain how you applied the process of elimination.

Exercise 7.5.

You are unsure about the answer in a multiple-choice math question but want to make an educated guess. The question is: "What is 5 + 3 × 2?" The available choices are:

1. **8**
2. **10**
3. **16**
4. **11**

Use guessing strategies to select the most reasonable option. Describe how you arrived at your answer.

Exercise 7.6.

After solving a question involving evaluating the function f(x) = 2x + 5 for x = 7, the following answer was obtained: "21".
Is this answer correct? If it is not, discuss what went wrong in the calculation and explain how you would review your work to find the correct answer.

1. **Correct it if necessary.**
2. **Show the steps involved.**

Answers with Explanations

Solution 7.4. Exercise 1:

We are tasked with finding the square root of 81. Let's apply the process of elimination.

- Option A: 7 is incorrect because $7 \times 7 = 49$, which is much less than 81.

- Option B: 9 seems reasonable because $9 \times 9 = 81$. This is a possible correct choice, so we keep it for now.

- Option C: 12 is too large since $12 \times 12 = 144$, which is much larger than 81.

- Option D: 10 is also too large because $10 \times 10 = 100$, which is larger than 81.

By eliminating the other options, the correct answer is clearly Option B, 9, because $9 \times 9 = 81$.

Solution 7.5. Exercise 2:

We need to evaluate $5 + 3 \times 2$. According to the order of operations (PEMDAS):

- First, we perform the multiplication: $3 \times 2 = 6$.

- Then, add $5 + 6 = 11$.

None of the options match the correct result directly, but it's easy to see that Option D (11) is the right choice if guessing out of uncertainty. In case of guessing, options involving much larger or smaller values can be eliminated.

Correct answer: **Option D.**

Solution 7.6. Exercise 3:

The problem asks us to evaluate the function $f(x) = 2x + 5$ when $x = 7$. The original answer given is "21", but let's see if that's accurate.
To start, substitute $x = 7$ into the function:

$$f(7) = 2(7) + 5 = 14 + 5 = 19$$

The correct answer is actually 19, not 21. The mistake may have occurred due to incorrect addition or oversight in calculation review.

Review Strategy:

- After plugging values into the function, always double-check the multiplication and addition steps.

- If there is uncertainty about your answer, try to substitute the result back into the problem definition to see if it makes sense relative to other given information.

Thus, the correct answer is 19.

"*John bought 3 boxes of apples. Each box contained 12 apples. He ate 2 apples from one box. How many apples does John have left in total? *"

At first glance, you might be tempted to multiply 3 by 12 and then subtract 2, assuming that John ate 2 apples from the total number of apples:

$3 \times 12 = 36$ and then $36 - 2 = 34$ (*incorrect answer*)

However, the trick here is in understanding the finer detail: John ate apples from only one box, not from all three. Correctly updating one box reduces the apples as follows:

$12 - 2 = 10$ (apples remaining in that box)

Thus, John's total remaining becomes:

$2 \times 12 + 10 = 24 + 10 = 34$ (*correct answer*)

The answer is indeed 34, but recognizing that the adjustment affects only one box is key.

Example 2: Extra Information

In some questions, extra numbers or details may be included to confuse you. Consider a question:

"A car travels 150 miles at 50 miles per hour and then it stops for lunch. If the lunch break lasts for 30 minutes, how long does it take for the car to reach its destination?"

The extra detail about the lunch break might catch your attention, but it is irrelevant. You only need the data from when the car is traveling.

Using the formula:

$$Time = \frac{Distance}{Speed} = \frac{150}{50} = 3\ hours$$

The lunch break does not change the time spent on the road, so the total time taken is still 3 hours.

7.3.2 Double-Checking Units

Another common mistake students make is overlooking units. Different units of measure (e.g., inches, feet, kilometers, hours, minutes) can complicate seemingly simple problems. Always check the units and convert them as necessary. Ignoring them can lead to wrong answers, especially in problems involving geometry or rates.

Example 3: Converting Units

Suppose you're asked to calculate the perimeter of a rectangle with side lengths provided in different units:

"One side is 3 feet, and the opposite side is 24 inches. What is the perimeter?"

First, you should recognize that inches and feet are different units. Converting 24 inches to feet, we get:

$$24\ inches = \frac{24}{12} = 2\ feet$$

Now calculate the perimeter. The formula for the perimeter P of a rectangle is:

$$P = 2 \times (\text{length} + \text{width})$$

Substituting the values:

$P = 2 \times (3\text{ feet} + 2\text{ feet}) = 2 \times 5 = 10\text{ feet}$ Thus, the correct perimeter is 10 feet.

7.3.3 Confirming Calculations

Lastly, it's essential to double-check your answers, especially in calculations that involve multiple steps or large numbers. Simple arithmetic errors can cause you to miss easy points. A great habit is to review the problem from start to finish once you arrive at an answer. Let's consider a few strategies for confirming calculations.

Example 4: Re-Verify with Estimation

Let's say you are adding several large numbers:

$$456 + 789 + 645$$

First, calculate the exact sum:

$$456 + 789 = 1245$$

$$1245 + 645 = 1890$$

The total is 1890. To avoid computation errors, it's a good idea to estimate the numbers by rounding them first:

$$456 \approx 500, \quad 789 \approx 800, \quad 645 \approx 600$$

The estimated sum becomes:

$$500 + 800 + 600 = 1900$$

Since your estimate (1900) is very close to the actual sum (1890), there's a good chance your original calculations are correct.

Example 5: Verification with Inverse Operations

You can also reverse your operations to check your results. Consider the equation:

$$5x + 7 = 27$$

Solving for x, first subtract 7 from both sides:

$$5x = 27 - 7 = 20$$

Then, divide by 5:

$$x = \frac{20}{5} = 4$$

To verify this, substitute $x = 4$ back into the original equation:

$$5(4) + 7 = 27 \text{ which simplifies to } \quad 20 + 7 = 27$$

Since both sides of the equation balance, your solution is confirmed to be correct.

7.3.4 Figures and Tables for Visual Assistance

To support these ideas, let's consider the following figure that illustrates unit conversion and a table that offers a handy reference for common unit conversions:

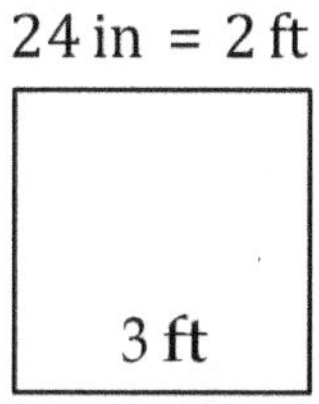

Rectangle dimensions

Figure 7.4: Visual representation of rectangle side lengths

Units	Conversion Factor
1 foot	12 inches
1 meter	100 centimeters
1 kilometer	1000 meters
1 hour	60 minutes
1 mile	5280 feet

Table 7.3: Common Unit Conversions

Remember, always be vigilant about units and double-check all calculations—both by reviewing your steps and through independent verification methods such as estimation or inverse operations.

By following these steps and practicing caution, you can avoid these common pitfalls and perform better on your GED Math test.

Exercises

Exercise 7.7.

1. Identify the error in the following question: "If you have 3 apples and you eat 2, how many apples do you have left?" What common pitfall should students be aware of in this type of question?

2. **Double-check the units in the following problem: "A car travels at 60 km per hour for 2 hours. What is the total distance in meters?" Describe the correct approach to ensure the units are consistent and correct.**

3. **Calculate the total for: $\frac{2}{5}+\frac{1}{3}-\frac{1}{2}$. Show all steps clearly, and confirm your final result is accurate by calculating again.**

Exercise 7.8.

Consider the following equation: $12+4\times(3-1)-10$.

1. **Solve the equation step by step.**
2. **Identify any potential pitfalls in performing operations in the wrong order.**

What key rule helps prevent making this mistake?

Exercise 7.9.

A recipe requires 2 cups of flour but you accidentally added 2 tablespoons more. A tablespoon is equivalent to 1/16 of a cup. Calculate the correct total amount of flour in cups. Did you accidentally add a significant extra amount? How does this example link to the common pitfall of mismanaging conversions between units?

Answers with Explanations

Solution 7.7.

The trick in the first question is subtle. The phrase "how many apples do you have left?" might lead to confusion. Students should focus on the actual arithmetic here, but it's important to be aware of potential language tricks in questions. This one is simple: the answer is 1 apple. Understanding the phrasing helps avoid unnecessary mistakes.

Solution 7.8.

The problem involves a units conversion issue. First, find the total distance in kilometers:

$$\text{distance} = 60\,\text{km/h} \times 2\,\text{h} = 120\,\text{km}$$

To convert kilometers to meters, use:

$$1\,\text{km} = 1000\,\text{m}$$

Thus:

$$\text{distance} = 120\,\text{km} \times 1000 = 120{,}000\,\text{m}$$

The common pitfall here is not converting units before answering the question. Make sure to check that the units in the question match the units in the answer.

Solution 7.9.
Calculate the sum:

$$\frac{2}{5} + \frac{1}{3} - \frac{1}{2}$$

Find a common denominator, which in this case is 30:

$$\frac{2}{50} = \frac{12}{30}, \quad \frac{1}{3} = \frac{10}{30}, \quad \frac{1}{2} = \frac{15}{30}$$

Now perform the operations:

$$\frac{12}{30} + \frac{10}{30} = \frac{22}{30}$$

Then subtract:

$$\frac{22}{30} - \frac{15}{30} = \frac{7}{30}$$

Final Result: $\frac{7}{30}$.

The key pitfall here lies in correctly managing the denominators and ensuring all calculations are accurate. Always recheck the arithmetic and reconsider the intermediate steps.

Solution 7.10.

Step-by-step solution of the equation $12 + 4 \times (3 - 1) - 10$:

$$\begin{aligned} &12 + 4 \times (3 - 1) - 10 = 12 + 4 \times 2 - 10 \\ &= 12 + 8 - 10 \\ &= 20 - 10 \\ &= 10 \end{aligned}$$

Pitfall: A common mistake would be to perform the operations in the wrong order. According to the order of operations (PEMDAS), you must handle parentheses first, then perform multiplication/division before moving to addition/subtraction. Skipping this rule leads to incorrect results.

Illustration:

PEMDAS steps

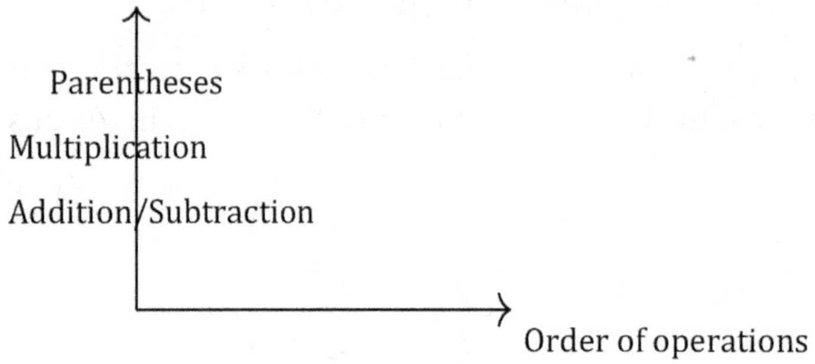

Solution 7.11.

The recipe calls for calculating how much flour was really added.

$$\text{Extra Flour} = 2\text{ tablespoons} = \frac{2}{16}\text{cups} = \frac{1}{8}\text{cups}$$

So, the total flour added is:

$$\text{Total flour} = 2\text{cups} + \frac{1}{8}\text{cup} = 2\frac{1}{8}\text{cups}$$

Pitfall: Mismanagement of unit conversions (e.g., tablespoons to cups) can cause errors. The extra amount was relatively small, only of a cup, but improper unit conversions could lead to larger errors in other situations. Clear understanding and double-checking are essential.

CHAPTER # 8

Tests

Math topics that you'll learn in this chapter:

- Exam
- Multiple-Choice Questions

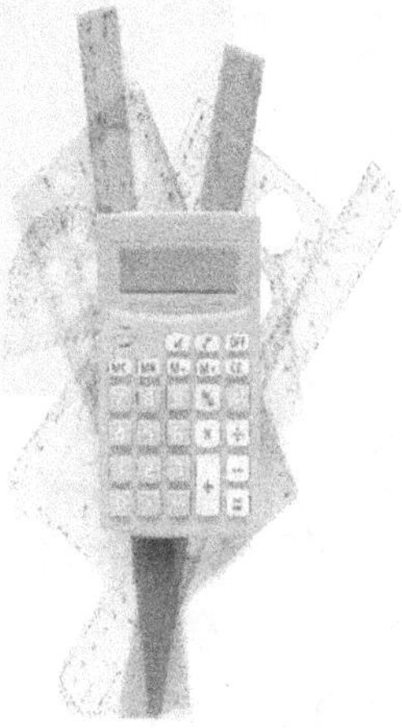

Scantron Answer Sheet for Exam

Student Name: ______________________________ Date: ______________

Instructions:

- Fill in the circle corresponding to your chosen answer for each question.
- Use a dark pen or pencil to mark your answers clearly.
- If you need to change an answer, erase or cross out your previous mark completely.

Question	A	B	C	D
1	○	○	○	○
2	○	○	○	○
3	○	○	○	○
4	○	○	○	○
5	○	○	○	○
6	○	○	○	○
7	○	○	○	○
8	○	○	○	○
9	○	○	○	○
10	○	○	○	○
11	○	○	○	○
12	○	○	○	○
13	○	○	○	○
14	○	○	○	○
15	○	○	○	○

Question	A	B	C	D
16	○	○	○	○
17	○	○	○	○
18	○	○	○	○
19	○	○	○	○
20	○	○	○	○

Note: Please ensure all selected answers are clearly marked. Review your answer sheet before submission to make sure you have answered all questions.

8.1 Exam

Multiple-Choice Questions

1. Which of the following represents the correct use of the parentheses in the order of operations?

A) 2 × (3 + 4) − 1 = 13

B) 2 × 3 + (4 − 1) = 11

C) 2 × 3 + 4 − 1 = 5

D) (2 × 3) + 4 − 1 = 1

2. If you multiply a fraction by its reciprocal, the result is:

A) 1

B) 0

C) The original fraction

D) Undefined

3. Solve for x in 3x − 5 = 16:

A) x = 5

B) x = 7

C) x = 3

D) x = 13

4. Which of the following is a prime number?

A) 9

B) 10

C) 11

D) 12

5. Simplify the following expression: 2(a − b) + 4a − 3

A) 6a − 2b − 3

B) 2a − 2b − 3

C) 6a − 2b + 3

D) $-2a + 2b - 3$

6. Solve $y - 3 = 2(y + 4)$:

A) $y = -11$

B) $y = -9$

C) $y = 9$

D) $y = 11$

7. Which of the following correctly describes a rational number?

A) A number that cannot be written as a fraction

B) A whole number

C) A number that can be expressed as a fraction of two integers

D) Any number less than zero

8. Simplify the expression $(2x)^2 + 3x \times 2x$:

A) $4x + 6x$

B) $4x^2 + 6x$

C) $2x + 3x$

D) $4x^2$

9. Which statement is true regarding the slope of the line $3x - 4y = 12$?

A) The slope is $\frac{4}{3}$

B) The slope is -3

C) The slope is 4

D) The slope is $\frac{3}{4}$

10. Solve the inequality: $2x - 3 < 9$

A) $x < 6$

B) $x > 6$

C) $x < 3$

D) $x > 3$

11. How do you calculate the area of a triangle?

A) $\frac{1}{2}$ × base × height

B) base × height

C) $\frac{base+height}{2}$

D) base + height

12. The formula for the volume of a cylinder is:

A) πr^2 + h

B) 2πr^2h

C) $\frac{4}{3}\pi r^3$

D) πr2h

13. Which integer is both a factor and a multiple of 3?

A) 6

B) 9

C) 1

D) 3

14. What is the greatest common factor (GCF) of 36 and 48?

A) 6

B) 12

C) 18

D) 2

15. What is the y-intercept of the equation y = 3x + 5?

A) 5

B) 3

C) −3

D) −5

16. Which of the following best represents a quadratic equation?

A) $y = 2x + 5$

B) $y = x^2 + 3x + 2$

C) $y = 3x + 6$

D) $y = -x - 4$

17. What is the probability of drawing an ace from a standard deck of 52 cards?

A) $\frac{1}{4}$

B) $\frac{1}{52}$

C) $\frac{4}{52}$

D) $\frac{13}{52}$

18. Evaluate the function $f(x) = 2x^2 - 3x + 1$ for $x = 2$:

A) $f(2) = 5$

B) $f(2) = 3$

C) $f(2) = 7$

D) $f(2) = 6$

19. What is the median of the data set {2, 4, 6, 8, 10}?

A) 4

B) 6

C) 5

D) 8

20. What is the difference between a scalene and an equilateral triangle?

A) Scalene has three sides of the same length, equilateral has different lengths.

B) Scalene has different lengths for all sides, equilateral has equal sides.

C) Scalene has two equal sides, equilateral has no equal sides.

D) Scalene does not exist, equilateral is the only type of triangle.

Detailed Explanations: Multiple-Choice Questions

1. Which of the following represents the correct use of parentheses in the order of operations?

Answer: A) $2 \times (3 + 4) - 1 = 13$

Explanation: The order of operations (PEMDAS) dictates that parentheses are evaluated first, then multiplication, and finally subtraction. In this case, $(3 + 4)$ is evaluated first, giving 7. Then, $2 \times 7 = 14$, and finally, $14 - 1 = 13$.

2. If you multiply a fraction by its reciprocal, the result is:

Answer: A) 1

Explanation: A reciprocal of a fraction is obtained by swapping its numerator and denominator.

For example, the reciprocal of 2/3 is 3/2. Multiplying a fraction by its reciprocal always results in 1.

3. Solve for x in $3x - 5 = 16$:

Answer: B) $x = 7$

Explanation: To solve for x, first add 5 to both sides of the equation: $3x = 21$. Then, divide both sides by 3: $x = 7$.

4. Which of the following is a prime number?

Answer: C) 11

Explanation: A prime number is a number greater than 1 that has no divisors other than 1 and itself. Among the given options, 11 is the only prime number.

5. Simplify the following expression: $2(a - b) + 4a - 3$

Answer: A) $6a - 2b - 3$

Explanation: Apply the distributive property: $2(a - b) = 2a - 2b$. Then, combine like terms: $2a + 4a - 2b - 3 = 6a - 2b - 3$.

6. Solve $y - 3 = 2(y + 4)$:

Answer: A) $y = -11$

Explanation: Distribute 2 to both terms inside the parentheses: $y - 3 = 2y + 8$. Subtract y from both sides: $-3 = y + 8$. Finally, subtract 8 from both sides: $y = -11$.

7. Which of the following correctly describes a rational number?

Answer: C) A number that can be expressed as a fraction of two integers

Explanation: A rational number is any number that can be written as a fraction, where both the numerator and the denominator are integers. Examples include 1/2, 3, or -7/8.

8. Simplify the expression $2x^2 + 3x \times 2x$:

Answer: B) 4x^2 + 6x

Explanation: First, square (2x): $2x^2 = 4x^2$. Then, multiply 3x by 2x: $3x \times 2x = 6x$. Now combine both terms: $4x^2 + 6x$.

9. Which statement is true regarding the slope of the line $3x - 4y = 12$?

Answer: B) The slope is -3

Explanation: To find the slope, first solve for y in terms of x:
$3x - 4y = 12$ becomes $y = 3/4x - 3$. The slope is the coefficient of x, which is 3/4.

10. Solve the inequality: $2x - 3 < 9$

Answer: A) $x < 6$

Explanation: First, add 3 to both sides: $2x < 12$. Then divide by 2: $x < 6$

11. How do you calculate the area of a triangle?

Answer: A) 1/2 × base × height

Explanation: The area of a triangle is given by the formula 1/2 × base × height.

12. The formula for the volume of a cylinder is:

Answer: D) πr^2h

Explanation: The volume of a cylinder is given by the formula V = πr^2h, where r is the radius and h is the height.

13. Which integer is both a factor and a multiple of 3?

Answer: D) 3

Explanation: 3 is a factor of itself and is also a multiple of 3.

14. What is the greatest common factor (GCF) of 36 and 48?

Answer: B) 12

Explanation: The GCF of 36 and 48 is the largest number that divides both, which is 12.

15. What is the y-intercept of the equation y = 3x + 5?

Answer: A) 5

Explanation: In the equation y = 3x + 5, the y-intercept is the constant term, which is 5.

16. Which of the following best represents a quadratic equation?

Answer: B) $y = x^2 + 3x + 2$

Explanation: A quadratic equation is one where the highest power of x is 2.

17. What is the probability of drawing an ace from a standard deck of 52 cards?

Answer: C) 4/52

Explanation: There are 4 aces in a deck of 52 cards, so the probability is 4/52.

18. Evaluate the function $f(x) = 2x^2 - 3x + 1$ for $x = 2$:

Answer: C) $f(2) = 7$

Explanation: Substitute $x = 2$ into the function: $f(2) = 2(2)^2 - 3(2) + 1 = 8 - 6 + 1 = 7$.

19. What is the median of the data set {2, 4, 6, 8, 10}?

Answer: B) 6

Explanation: The median is the middle number when the data set is arranged in order. In this case, it is 6.

20. What is the difference between a scalene and an equilateral triangle?

Answer: B) Scalene has different lengths for all sides, equilateral has equal sides.

Explanation: A scalene triangle has no equal sides, while an equilateral triangle has all sides equal.

Scantron Answer Sheet for Final Exam

Student Name: ______________________ Date: __________

Instructions:

- Mark one answer for each question.
- Completely fill the circle with corresponding answer.
- Use only a pencil or blue/black pen.

Question	A	B	C	D
1	○	○	○	○
2	○	○	○	○
3	○	○	○	○
4	○	○	○	○
5	○	○	○	○
6	○	○	○	○
7	○	○	○	○
8	○	○	○	○
9	○	○	○	○
10	○	○	○	○
11	○	○	○	○
12	○	○	○	○
13	○	○	○	○
14	○	○	○	○
15	○	○	○	○

Question	A	B	C	D
16	○	○	○	○
17	○	○	○	○
18	○	○	○	○
19	○	○	○	○
20	○	○	○	○

Note: Please ensure all selected answers are clearly marked. Review your answer sheet before submission to make sure you have answered all questions.

8.2 Final Exam

Multiple-Choice Questions

1. Which of the following is a solution to the equation $5x - 2 = 13$?

A) $x = 1$

B) $x = 3$

C) $x = 5$

D) $x = 6$

2. If $y = 3$, what is the value of $2y + 4$?

A) 5

B) 7

C) 8

D) 10

3. What is the value of x in the equation $3x + 6 = 21$?

A) 4

B) 5

C) 6

D) 7

4. What is the value of y in the equation $4y - 5 = 7$?

A) 4

B) 3

C) 2

D) 1

5. Which is not a property of an equilateral triangle?

A) All angles are equal.

B) The triangle has rotational symmetry.

C) All sides are of equal length.

D) Two sides are longer than the third.

6. Which of the following shapes has exactly one pair of parallel sides?

A) Square

B) Rhombus

C) Trapezoid

D) Rectangle

7. Solve for x: $2x + 4 = 16$.

A) $x = 4$

B) $x = 6$

C) $x = 3$

D) $x = 5$

8. Which operation would you perform first in the expression 3 + 4 × 2 following the order of operations (PEMDAS)?

A) Add 3 and 4

B) Multiply 4 and 2

C) Subtract 4 from 2

D) Multiply 3 and 4

9. What is the mean of the data set: 4, 8, 12, 16, 20?

A) 12

B) 10

C) 14

D) 16

10. A function f(x) is defined as f(x) = 2x + 1. What is f (3)?

A) 5

B) 7

C) 9

D) 11

11. Solve the inequality: $3x - 2 < 4$.

A) $x < 2$

B) $x < 3$

C) $x < 4$

D) $x < 6$

12. Convert 0.75 into a fraction.

A) $\frac{1}{2}$

B) $\frac{3}{4}$

C) $\frac{7}{10}$

D) $\frac{5}{6}$

13. Which formula represents the area A of a circle with radius r?

A) $A = \pi r^2$

B) $A = 2\pi r$

C) $A = \frac{\pi r^2}{2}$

D) $A = \frac{1}{2}\pi r^2$

14. Find the slope of a line passing through the points (2, 3) and (4, 7).

A) $\frac{1}{2}$

B) 1

C) 2

D) 3

15. Solve the equation: $x^2 - 9 = 0$.

A) $x = 0$

B) $x = 3$

C) $x = \pm 3$

D) $x = -3$

16. Factor the expression: $x^2 + 5x + 6$.

A) $(x + 2)(x + 3)$

B) $(x + 1)(x + 6)$

C) $(x - 2)(x + 3)$

D) $(x + 2)(x - 6)$

17. The point (2, −5) lies in which quadrant of the Cartesian plane?

A) First quadrant

B) Second quadrant

C) Third quadrant

D) Fourth quadrant

18. What is 60% of 150?

A) 90

B) 75

C) 45

D) 30

19. In a right triangle, if one angle is 90°, what is the sum of the other two angles?

A) 45°

B) 60°

C) 90°

D) 180°

20. If the perimeter of a rectangle is 24 and the width is 4, what is the length?

A) 5

B) 6

C) 8

D) 10

Detailed Answer Sheet: Multiple-Choice Questions

1. Which of the following is a solution to the equation 5x - 2 = 13?

Answer: B) x = 3

Explanation: To solve for x, add 2 to both sides of the equation: 5x = 15. Then, divide both sides by 5: x = 3.

2. If y = 3, what is the value of 2y + 4?

Answer: D) 10

Explanation: Substitute $y = 3$ into the expression: $2(3) + 4 = 6 + 4 = 10$.

3. What is the value of x in the equation $3x + 6 = 21$?

Answer: B) $x = 5$

Explanation: Subtract 6 from both sides: $3x = 15$. Then divide both sides by 3: $x = 5$.

4. What is the value of y in the equation $4y - 5 = 7$?

Answer: A) $y = 4$

Explanation: Add 5 to both sides: $4y = 12$. Then divide both sides by 4: $y = 4$.

5. Which is not a property of an equilateral triangle?

Answer: D) Two sides are longer than the third.

Explanation: An equilateral triangle has all sides of equal length, and all angles are equal. The statement that "two sides are longer than the third" is false.

6. Which of the following shapes has exactly one pair of parallel sides?

Answer: C) Trapezoid

Explanation: A trapezoid is defined as a quadrilateral with exactly one pair of parallel sides.

7. Solve for x: $2x + 4 = 16$.

Answer: A) $x = 4$

Explanation: Subtract 4 from both sides: $2x = 12$. Then divide both sides by 2: $x = 4$.

8. Which operation would you perform first in the expression $3 + 4 \times 2$ following the order of operations (PEMDAS)?

Answer: B) Multiply 4 and 2

Explanation: According to PEMDAS, multiplication comes before addition. Therefore, 4×2 is performed first.

9. What is the mean of the data set: 4, 8, 12, 16, 20?

Answer: A) 12

Explanation: The mean is the sum of the values divided by the number of values. $(4 + 8 + 12 + 16 + 20) / 5 = 60 / 5 = 12$.

10. A function f(x) is defined as $f(x) = 2x + 1$. What is f (3)?

Answer: B) 7

Explanation: Substitute $x = 3$ into the function: $f(3) = 2(3) + 1 = 6 + 1 = 7$.

11. Solve the inequality: $3x - 2 < 4$.

Answer: A) $x < 2$

Explanation: Add 2 to both sides: $3x < 6$. Then divide by 3: $x < 2$.

12. Convert 0.75 into a fraction.

Answer: B) 3/4

Explanation: 0.75 is equivalent to 75/100, which simplifies to 3/4.

13. Which formula represents the area A of a circle with radius r?

Answer: A) $A = \pi r^2$

Explanation: The area of a circle is given by the formula A = = πr^2, where r is the radius.

14. Find the slope of a line passing through the points (2, 3) and (4, 7).

Answer: C) 2

Explanation: The slope of a line is calculated as (y2 - y1) / (x2 - x1). For the points (2, 3) and (4, 7), the slope is (7 - 3) / (4 - 2) = 4 / 2 = 2.

15. Solve the equation: $x^2 - 9 = 0$.

Answer: C) x = ±3

Explanation: Add 9 to both sides: $x^2 = 9$. Then take the square root of both sides: x = ±3.

16. Factor the expression: $x^2 + 5x + 6$.

Answer: A) (x + 2) (x + 3)

Explanation: The factors of 6 that add up to 5 are 2 and 3. Therefore, the factored form is (x + 2) (x + 3).

17. The point (2, -5) lies in which quadrant of the Cartesian plane?

Answer: D) Fourth quadrant

Explanation: In the fourth quadrant, x is positive, and y is negative. The point (2, -5) fits this description.

18. What is 60% of 150?

Answer: A) 90

Explanation: To find 60% of 150, multiply 150 by 0.60: 150 × 0.60 = 90.

19. In a right triangle, if one angle is 90°, what is the sum of the other two angles?

Answer: C) 90°

Explanation: The sum of all angles in any triangle is 180°. If one angle is 90°, the sum of the other two angles must be 90°.

20. If the perimeter of a rectangle is 24 and the width is 4, what is the length?

Answer: B) 6

Explanation: The perimeter of a rectangle is calculated as 2(length + width). Given the width of 4, the equation becomes 24 = 2(length + 4). Solving for length gives length = 6

Author's Final Note

GED Math Practice 2026 was prepared to help learners build confidence with the mathematics skills tested on the GED Mathematical Reasoning exam.

Use the lessons, practice questions, and answer explanations as a steady review path. When a problem is difficult, slow down, write out each step, and return to the related topic before moving ahead.

This workbook is designed for adult learners, high school equivalency students, tutors, and independent study. The goal is practical progress: stronger number sense, clearer algebra skills, better geometry reasoning, and more confidence with GED-style word problems.

Thank you for studying with Mathlibros. Keep practicing, check your work carefully, and let each solved problem become one more step toward your goal.

www.ingramcontent.com/pod-product-compliance
Lightning Source LLC
LaVergne TN
LVHW060546061126
851857LV00017B/1915